飧宴
Feast

蒋国兴作品集
Jiang Guoxing Portfolio

主编 蒋国兴　策划 徐宾宾

江苏科学技术出版社

Preface

Xu Pin Design Company catches up with the trend and is ready to issue a book with great care. Before writing this article, I cannot help thinking a lot of things. Our company was established in 2006, and there're only three employees at that time. Now more than 40 people are in our team after 8 years' development. We have gone through a lot, while we never change design principle of our company, that is, to "only create original design". The communication with the proprietor during the process is most touching, I don't know what to tell you, but only say something usually told to the proprietors.

Design, in fact, is a process of being in love with proprietor. Most of the time, the communication between the owner and the designer is key to design. During this period, the designer keeps communicating with the proprietor, and considers requirement and favorite of every customer carefully. He should respect customers when focusing on his design. It's very important for both designer and proprietor to consider for each other. Regarding himself as a true friend of the proprietor, guessing what the proprietor requests from this aspect, the designer should be considerate and provide sincere and practical suggestions. Anyway, there're some occasions when the designer holds different opinions from the proprietor, and even debate happens. After all, the proprietor is not professional and possesses insufficient knowledge of design. Meanwhile, designer pays more attention to the case and wants it perfect. The unavoidable insist is based on the responsibility and protection for the case. The aim of the designer is to perfect the case, which is exactly what the proprietor is looking forward to.

Therefore, in modern business society, that whether the design of the commercial space can benefit for the proprietor is the crucial point of design. And the value is mostly decided by the effect of comprehensive ability of the propriety himself. The choice of location, field visit, product positioning and executive ability will affect the case.

The supports from proprietors contribute to all successful cases done by Xu Pin in these years to. Thanks for all those supports and supporters, then we can present the most sincere works to the public. We dedicate this book to those who know, like, care about, and love Xu Pin.

Jiang Guoxing

序言

叙品设计赶了潮流，小心翼翼地准备出书了。拿起笔，思绪万千，从2006年到如今，8年时间尝尽酸甜苦辣，公司从最早的3个人发展到现在的40多个人，我们经历了很多，但公司的设计原则从未改变过，秉持"只做原创设计"。过程中与业主的交流最令叙品感动，不知道要写什么给大家，只是把平时同业主沟通时的几句话，说说罢了。

设计，其实是一个与业主"恋爱"的过程，很多时候业主与设计师的沟通是设计的关键。在这个过程中，设计师反复与业主沟通交流，设计师要对每位客户的需求、喜好认真考量。在尊重自己作品的同时，也尊重客户。无论是设计师还是客户，换位思考都是很重要的。把自己作为客户的知心朋友来看待，从这个角度来揣摩客户心理，想客户所想，提供切实真诚的建议。然而有时设计师会坚持己见，与业主意见不合甚至激烈争论。毕竟，业主不是内行，对设计方面的工作知道的比较少，同时设计师对案子重视，想令其结果完美，不得已的坚持，其实是对案子的负责与爱惜，其初衷与业主相同，让案子更接近完美。

那么在现代商业社会里，商业空间的设计，对于业主而言，让其盈利才是这个设计最重要的。这往往取决于业主的综合能力所带来的影响。业主前期对案子的地点的选择、实地的考察、产品的定位把关，以及实施能力，这些都能影响案子的成功与否。

叙品这些年做过许多成功的案子，这与业主的支持是分不开的，感谢他们支持叙品，叙品才能呈现最诚挚的作品给众人，谨以此书献给所有认识叙品，喜欢叙品，关注叙品，关爱叙品的人。

蒋国兴
2014年3月

目录 Contents

Yiyang Impression Cafe 一阳咖啡印象店	006
016	French Restaurant of Yiyang Impression 一阳印象法式西餐厅
Yiyang Brazilian Barbecue 一阳巴西烧烤店	022
028	Yiyang Minzhu 一阳咖啡民主店
Blanca 布蓝卡	036
046	Xupin Café 叙品咖啡
Sakata Japanese Restaurant Phase I SAKATA酒田日本料理一期	056
066	Sakata Japanese Restaurant Phase II SAKATA酒田日本料理二期
Suzhou Sakata Japanese Cuisine 苏州酒田日本料理	076
086	9.9inPizza Italian Restaurant 9.9inPizza意式餐厅
Tangcheng Hot Pot City 汤城火锅店	094
102	Xupin Design Company Bar (Kunshan) 叙品设计公司酒吧（昆山）
Xinjiang Huazhi Boiling Fish 新疆花枝沸腾鱼	112

122	Refined Food Restaurant 精膳
132	Menyingtianxia Hot-Pot Restaurant 门迎天下火锅店
144	Bamboo Stream No. 1 Restaurant 竹溪一号
154	Mashijiu Pot Restaurant on Danlu Road 马仕玖煲丹露店
162	Yuanshan I 原膳一期
182	Yuanshan II 原膳二期
196	Shenhai Yihao 深海壹號
210	Mashijiu Pot-Aksu Restaurant 马仕玖煲阿克苏店
224	Oulandi Café 欧兰迪咖啡厅
240	Eastern Fence Narration 东篱·叙
256	Xupin Design Xinjiang Branch Office (Times Square) 叙品设计新疆分公司（时代广场）
270	Xupin Design Shandong Branch 叙品设计山东分公司
280	Xupin Design (Zhenchuan Road) 叙品设计（震川路店）
292	Longhai Construction 龙海建工

Yiyang Impression Cafe

一阳咖啡印象店

Designer: Jiang Guoxing
Design Company: Xupin Design Decoration Engineering Co., Ltd.
Building Area: 1,580 m²
Main Materials: Wood flooring, gold foil, brown textured marble, mosaic, etc.

设 计 师：蒋国兴
设计公司：苏州叙品设计装饰工程有限公司
建筑面积：1580平方米
主要材料：实木地板、金箔、啡网纹大理石、马赛克等

In the case, the main tone of orange is matched with soft lighting and the furniture of the nostalgic style, very cozy and comfortable. The arc-shaped design of chairs and lamps is very unique, and such a design well saves the interior space, perfectly integrating with the overall style of the cafe.

As for the cellar, it is very transparent and bright. A wall of glass cabinet is filled with various wines of different age, very imposing and attractive.

The finest place is a showcase filled with a variety of refined products outside the washing room. It is dazzling and very chic, which is the imgenuity of this cafe.

设计师运用的色彩很柔和，以橘黄色为咖啡厅的主色调，略带怀旧风格的家具，再配上柔和的灯光，十分温馨舒适。椅子和灯的圆弧形设计很特别，椅子这样设计最节省空间，和咖啡厅的设计风格浑然天成地融合在一起。

酒窖则设计得十分通透明亮。满墙的玻璃酒柜，装满了各式各样的不同年份的红酒，远远望去很有气势，也很独特。

最别致的是，设计师在洗手间外围放置了一个装满各式精品的陈列柜，琳琅满目，却又如此高雅，这就是这个咖啡厅的高明之处。

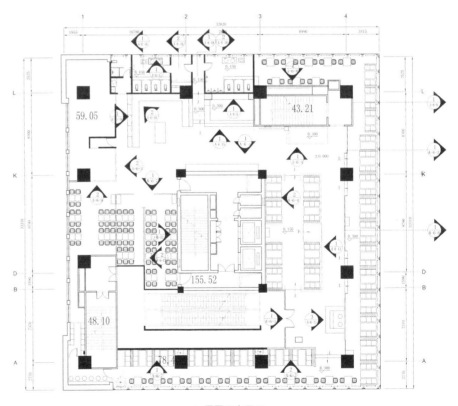

二层平面布置图

011

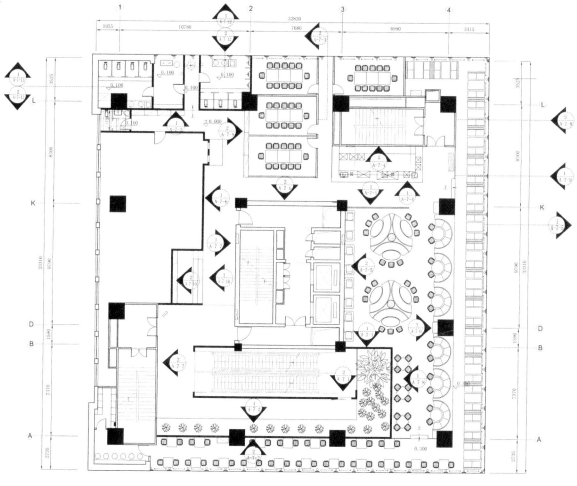

三层平面布置图

French Restaurant of Yiyang Impression

一阳印象法式西餐厅

Designer: Jiang Guoxing
Design Company: Xupin Design Decoration Engineering Co., Ltd.
Project Location: Urumqi, Xinjiang
Building Area: 1,100 m²
Main Materials: Translucent marble, dark wood flooring, dark wood veneer, dark brown textured marble, blue wallpaper, etc.

设 计 师：蒋国兴
设计公司：苏州叙品设计装饰工程有限公司
项目地点：新疆乌鲁木齐
建筑面积：1 100平方米
主要材料：透光云石、深色实木地板、深色木饰面、深啡网纹大理石、湖蓝色壁纸等

With the favorable geographical location, Yiyang Impression is the first French restaurant in Urumqi specializing in French cuisine, located in the downtown of Urumqi in Xinjiang.

In the case, the designer not only demonstrates the nobleness and elegance of the traditional French restaurant, but creates a warm and pleasant restaurant with the casual atmosphere. The main tone of orange and brown is matched with soft lighting and the furniture of the nostalgic style, very cozy and comfortable. Besides, the arc-shaped design of chairs and lamps is very unique, which is not beautiful but saves the interior space, perfectly integrating with the overall style of the restaurant.

As for the cellar, it is very transparent and bright. A wall of glass gradevin is filled with various wines of different age which exude an atmosphere of history, very unique and attractive, so that the customers unconsciously take a seat and taste wine with friends.

The finest place is a showcase filled with a variety of refined products outside the washing room. It is dazzling, giving a visual surprise.

一阳印象店是一家法式西餐厅，位于新疆乌鲁木齐繁华街区，地理位置得天独厚，是乌鲁木齐首家专营法国菜的西餐厅。

设计师不仅表现出传统法式餐厅的高贵典雅，更以休闲氛围融合主情调，创造一个温暖怡人的法式西餐厅。设计师运用的色彩柔和，以橘黄色和咖啡色为西餐厅的主色调，略带怀旧风格的家具，再配上柔和的灯光，十分温馨舒适。椅子和灯的圆弧形设计很特别，不仅美观，也节省室内空间，动线流畅，与西餐厅的整体风格浑然天成地融合在一起。

酒窖则设计得十分通透明亮。满墙的玻璃酒柜，装满了各式各样不同年份的红酒，散发出历史的气息，远远望去很独特，让人不经意间入座，品一品红酒与好友畅谈。

最别致的是设计师在洗手间外围放置了一个装满各式精品的陈列柜，琳琅满目，在视觉上给人惊喜。

Yiyang Brazilian Barbecue

一阳巴西烧烤店

Designer: Jiang Guoxing
Design Company: Xupin Design Decoration Engineering Co., Ltd.
Project Location: Urumqi, Xinjiang
Building Area: 1,100㎡
Main Materials: Brown wooden flooring, light wood veneer, copperlite glazing, etc.

设 计 师：蒋国兴
设计公司：苏州叙品设计装饰工程有限公司
项目地点：新疆乌鲁木齐
建筑面积：1 100平方米
主要材料：咖啡色实木地板、浅色木饰面、铜条玻璃等

Yiyang Brazilian Barbecue is a branch of Yiyang coffee, located in the downtown of Urumqi. The owner introduced Brazilian barbecue which is well-known by people of Xinjiang according to the local characteristics and habits. Brazilian barbeque is the Brazilian state banquet and loved by South American countries. Through the evolution and transmission of ages, Brazilian barbecue is now expanding rapidly in major Chinese cities and has swept through China, warmly welcomed by Chinese.

一阳巴西烧烤店是一阳咖啡旗下的一个分店，位于乌鲁木齐市中心。业主针对新疆本土特色、生活习惯，引进了对于新疆来说并不陌生的巴西烧烤。巴西烧烤是巴西的国宴，深受南美国家的喜爱，经过历史的演变、传承，到现在巴西烧烤在中国的各大城市迅速扩张，席卷了大江南北，令许多国人也对它喜爱不已。

In the case, the designer takes the style of romance and leisure, using brown as the main color and dark red as the secondary color in the dining hall. The seating arrangement is orderly and makes a rational use of the space so that the interior space has clear and bright moving lines, increasing the comfort of the space. Besides, the seats are well complementary with the dark red blocks in the ceiling. At the entrance, the dark red curtain at the top is matched with the hazy lighting. The wooden veneer of aging treatment is loaded with decorative paintings, and fine copperlite glazing brings a chic romance. Next to the booths in the hall is a row of French windows with the curtain of brown and purple-red. The warm colors make the entire space filled with passion and vitality in the romantic atmosphere, presenting a special charm. Meanwhile, the French window increases the indoor lighting and brings more enjoyment for customers. Besides, a row of buffet cabines are arranged in a corner of the hall in order to continue the characteristic of Brazilian barbecue—taking some buffet before the barbecue. The wall decoration of irregular paintings makes this space have a humanistic temperament and the beautiful appearance.

Such a romantic and comfortable space is filled with Brazilian flavor. Friends getting together here can enjoy the sincere friendship and the real life.

本案设计师以浪漫休闲为主格调，咖啡色为餐厅的主色，暗红色为辅色。散座分布井然有序，合理运用空间，使内部空间动线清晰明朗，增加空间的舒适度，与方块整列的暗红顶上下呼应，尤显特别。入口处顶面暗红色纱幔，配以迷离的灯光，做旧处理的木饰面，记录过往的装饰画，精致的铜艺玻璃，营造出浓郁的浪漫气息。餐厅沿边卡台旁是一排落地窗，咖啡色和紫红色相间的窗帘，以热情的色彩使整个空间浪漫中带着热情和活力，两相结合别有韵味，也增加了室内的采光，增添了更多的观赏性。餐厅的另一角是一排自助餐柜，延续了巴西烧烤的特色吃法，在烤肉上来之前客人会先吃一些自助小餐。设计师以不规则挂画作墙面装饰，使空间极具人性化的同时也极具美感。

闲暇之余，在此热情、活跃、浪漫的舒适空间，享受着鲜美粗犷带着芬芳的巴西风味，感受着朋友间真挚的友谊，以及浓郁的氛围，生活不应该就是这样的吗？

Yiyang Minzhu

一阳咖啡民主店

Designer: Jiang Guoxing
Design Company: Xupin Design Decoration Engineering Co., Ltd.
Project Location: Urumqi, Xinjiang
Building Area: 960 ㎡
Main Materials: Plywood, wooden flooring, black paint, tawny mirror, etc.

设 计 师：蒋国兴
设计公司：苏州叙品设计装饰工程有限公司
项目地点：新疆乌鲁木齐
建筑面积：960平方米
主要材料：复合板、实木地板、黑色漆、茶镜等

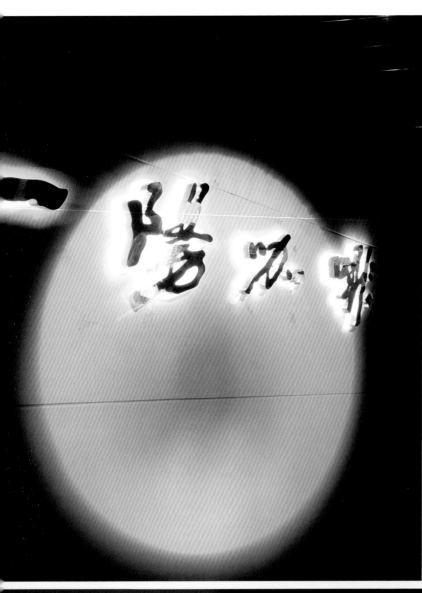

Nostalgic and romance are not the only choice of café any more.

The entrance hallway is unique. The waving background wall of stone pillars looks like lettering panel of ancient printing with rich culture. The chairs of coffee area in the traditional shape of arc, while the color of the chairs is very eye-catching: bright rosy red and dark red, which makes the chairs seem very fashionable.

怀旧浪漫，不再是咖啡厅的唯一选择。

咖啡厅的入口走道别具一格，一排高低起伏不平的石柱背景墙，宛如古人的印刷术中的字版，具有浓郁的文化气息。咖啡区的椅子选用了传统的半圆弧形，但椅子的颜色却很跳跃，鲜艳的玫红色和大红色交错其中，显得时尚、前卫。

Different from other normal café, even its tiny details here such as the layout of lighting and colors, modern style of wine cellar present a new style—the perfect combination of fashion and classic. The comparison between the virtual and real, bright and dark combines the traditional spirit and modern sense. Large areas using small sized dull polished red bricks complemented by home-made stainless steel tube lightings bring visual impact in the natural and friendly surroundings. The balance of the original and modern highlights this café.

A large area of French window is adopted in tea art area. And there're Chinese traditional wooden lattice grilles to complement it. In the sunshine and light, the lattice will leave shadows in the space dimly. The case is low-key, but exquisite in details, and as delicate and fragrant as tea.

　　与一般的咖啡厅不同的是，这里，即使是小小的细节，如灯光色彩的布置、酒窖的现代风格设计，都强烈地体现出一种新风格——时尚和古典的完美融合。虚与实、明与暗的对比实现一种传统精神和现代感的融合。空间中大面积采用特殊小规格磨砂面红砖，再辅以自制不锈钢管灯饰，在自然、亲切的环境中形成视觉冲击，原始与现代的平衡成为店内的一大亮点。

　　茶艺区有大面积的落地玻璃采光墙，配以中式传统木格窗花，在日光及户外灯光的照射下，木格隐隐约约投影在空间室内，不事张扬。细部精致的创意设计，如茶香般细腻，幽香致远。

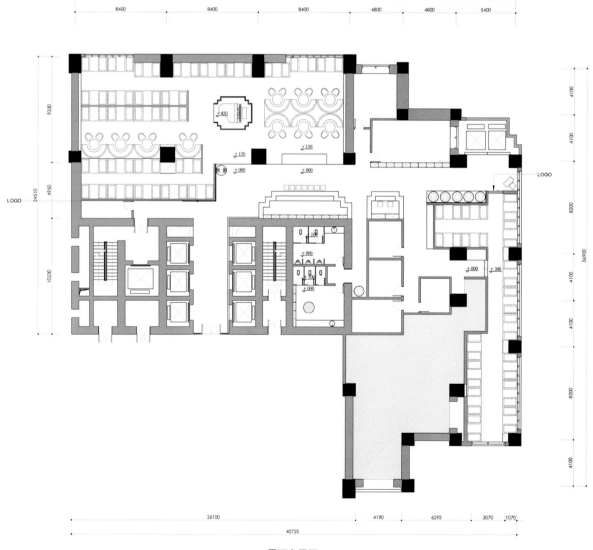

平面布置图

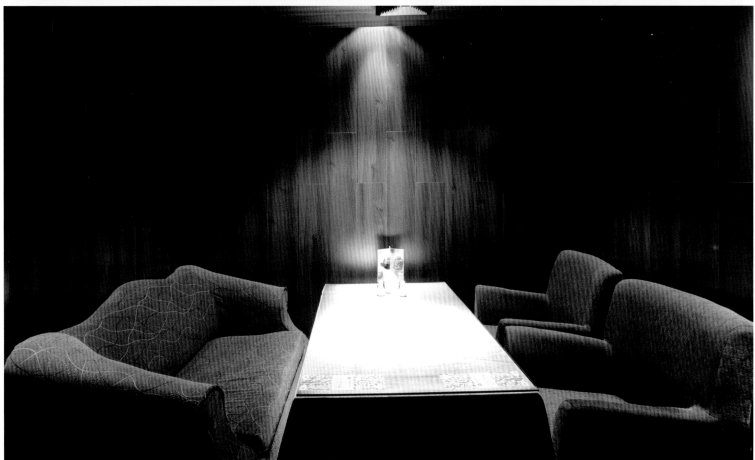

Blanca

布蓝卡

Designer: Jiang Guoxing
Design Company: Xupin Design Decoration Engineering Co., Ltd.
Project Location: Urumqi, Xinjiang
Building Area: 1,200 ㎡
Main Materials: Stainless steel, tawny mirror, wood flooring, foils, etc.

设 计 师：蒋国兴
设计公司：苏州叙品设计装饰工程有限公司
项目地点：新疆乌鲁木齐
建筑面积：1 200平方米
主要材料：不锈钢、茶镜、木地板、金箔等

At the beginning of the design, the designer refreshes the understanding of the coffee culture.

The space adheres to the theme of returning to the nature and introduces the pastoral elements with the use of green and yellow to create a natural atmosphere, integrating the elements of various famous coffees into the refreshing music so as to bring people a relaxed mind. All the materials, symbols, lighting, color, and the shape are the expression of the theme.

The jungle landscape in a corner of the hall directly attracts people into the theme with the gurgling water and rich fragrance, warm and relaxed, which allows people to get a cozy and relaxed experience from the visual, auditory, tactile and gustatory means in a short stay so as to forget the hubbub and achieve a physical and mental tranquility. The large use of wood, rattan, animal arnament, and gray-green wallpaper of leaves pattern makes the theme run through the entire space, aiming to create a quiet and tranquil atmosphere out of the world.

The project finally shows a pure and clean feeling. The essence of the original design is to create a new design in the original thinking and understanding of the project, accomplishing the breakthrough and an international perspective.

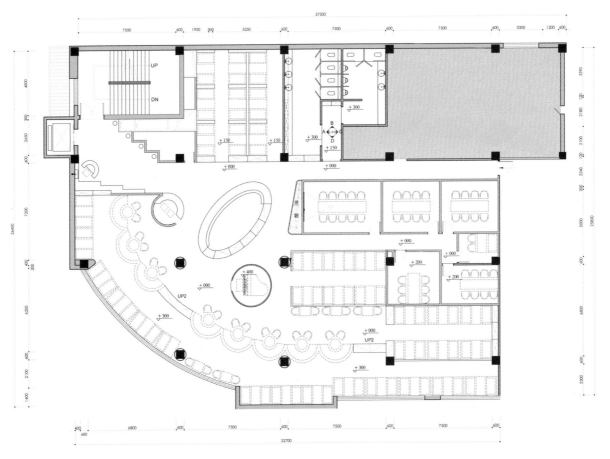

平面布置图

设计师在设计之初重新梳理了对咖啡文化的理解。

空间秉承了回归自然的主题，将田园的元素引入其中，使用大量的绿色与黄色，营造如自然的氛围，在沁人心脾的灵性音乐中糅入各种特色咖啡，让人的心灵充分放松和回归。所有的设计素材、符号、灯光、色彩、形体，均是为表达这个概念。

大厅一角的丛林景观将人们直接引入主题，当中流水潺潺，暖香扑鼻，温暖而放松，让人们在短短的逗留当中尽得视觉、听觉、触觉、味觉的奇妙体验，达到身心的宁静，忘却尘嚣。大量采用木、藤、动物饰品、灰绿色叶片图案墙纸，以回归自然的主题始终贯穿整个空间，旨在营造一种脱离尘世的静谧氛围。

项目最终呈现的是一种从未有过的纯粹与干净。原创设计的本质是原创的思维体系与原创性地理解项目，从而塑造全新的设计语言，获得具有国际化视界的可能性与突破性。

Xupin Café

叙品咖啡

Designer: Jiang Guoxing
Design Company: Xupin Design Decoration Engineering Co., Ltd.
Project Location: Kunshan, Jiangsu
Building Area: 1,440 m²
Main Materials: Plywood plate, wooden flooring, black paint, gold foil, etc.

设 计 师：蒋国兴
设计公司：苏州叙品设计装饰工程有限公司
项目地点：江苏昆山
建筑面积：1 440平方米
主要材料：复合板、实木地板、黑色漆、金箔等

Some people release passion through coffee, some express lovesickness through coffee, some get inspiration from coffee, some can relax themselves by coffee, some refresh themselves by coffee and some take some comfort from coffee... At leisure time, people go to café to review their lives.

This case is located in the most prosperous Ren Min Road in downtown of Kunshan city. It possesses excellent position and open layout, so that you can experience the prosperity of the city while sitting quietly here.

The atmosphere is tranquil, elegant and exquisite. Designers use modern methods to express the elements of environment by avoiding noise and reducing a little bit luxury, which combines food and culture. Indoor decoration elements present other images. A lot of wooden blocks are piled at the entrance as a decorative wall, providing a pure background for the whole surroundings which leaves a lot of space for people to imagine. The bar counter, cashier counter and decorative lighting combine perfectly and set simple style for the restaurant. Gold foil are used properly to show the noble but not luxurious space, just like telling people inside stories with rich and decent content, starting a journey both for heart and delicious foods.

From the aspect of details, various kinds of materials can offer perfect touching experience. When the warm light is on the brown floor, a dreamy atmosphere will be created. And modern, simple and fashion sofas can even reflect freshness. In fact, the fashionable and modern Xupin Café considers the customers' needs and expresses the care in every detail, which is approved by customers and attracts a lot of people to enjoy inside.

有人用咖啡释放激情，有人用咖啡寄托相思，有人用咖啡激发灵感，有人用咖啡放松心情，有人用咖啡驱除疲乏，有人用咖啡抚慰心伤……闲暇时，走进咖啡店，一杯咖啡叙品人生百味。

本店位于昆山市中心最繁华的街道——人民路上，拥有极为优越的地理位置及宽敞的布局，让顾客尽揽都市繁华，坐享静谧天地。

在环境气氛的营造上，为营造几分安静、雅致，回避喧嚣，减去一分奢华，将餐饮与文化融合在一起，用现代化的手法表现环境氛围。室内景观装饰元素展现出另一种气象，入口处大面积的木块叠加装饰墙为环境提供了一个很纯粹的背景，给人以遐想的空间，让吧台、收银台以及灯饰完美结合，奠定了餐厅的现代简约风格。适度使用金箔饰面，渲染空间环境的尊贵，但不奢华，传递出的是几分高贵，为来这里的人们开启一场属于心灵、味蕾的双重之旅。

在细节处理上，许多变化多端的材料给人触觉上的完美体验。当温馨、和煦的光线洒在棕色的地板上时，散发出一种浪漫的气息，而且现代、简洁、时尚的沙发椅更散发出一种清新。时尚现代的叙品咖啡结合消费者的心理诉求，将设计理念渗透到每个细节，深得消费者认同，客流量剧增。

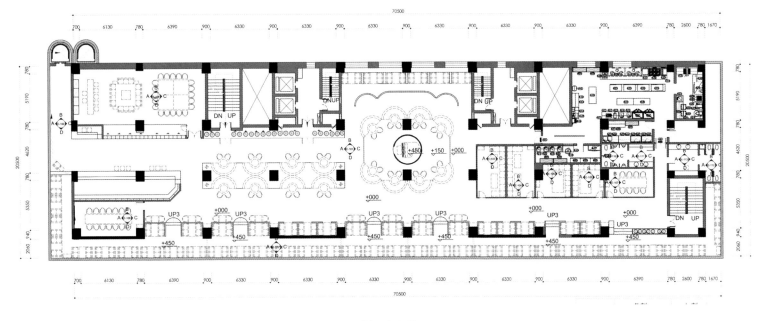

平面布置图

Sakata Japanese Restaurant Phase I

SAKATA酒田日本料理一期

Designer: Jiang Guoxing
Design Company: Xupin Design Decoration Engineering Co., Ltd.
Project Location: Kunshan, Jiangsu
Building Area: 380 ㎡
Main Materials: Ebony, Chinese black brick, woven glass, hair stone, stainless steel tube, etc.

设 计 师：蒋国兴
设计公司：苏州叙品设计装饰工程有限公司
项目地点：江苏昆山
建筑面积：380平方米
主要材料：黑檀木、中国黑砖、布纹玻璃、毛面锈石、不锈钢管等

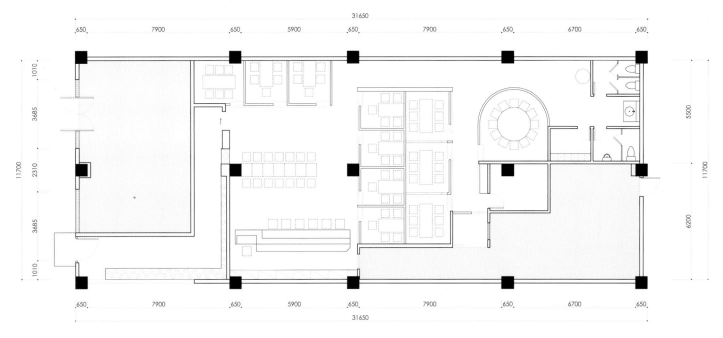

平面布置图

Owners expect designers can retain the exquisite taste of Japanese culture and at the same time, make innovation in the design. Therefore, the design concept of the case is identified as fashion, personality, and not losing the strict Japanese restaurant culture.

We keep the characteristics of chastity in Japanese culture in the use of materials, such as natural hair stone and rough granite, and some fashion materials such as mirror, mosaic. Interior design reflects the thinking of the transition from the old to the new, integration of Japanese and western style, and is fashionable, free, to create a diversified diet space. Entrance landscape corridor is the inspiration from a Japanese well-known folk song "spring in the north" to employ the white birch wood. Landscape design not only attracts in-store customers, for people outside the shop which also can create a strong visual impact.

业主期望设计师在设计时能保留日本文化的精致品味的同时，更能有所创新。因此，本案的设计理念确定为体现时尚的、潮流的、个性的，又不失严谨的日本餐厅。

在材料的运用上，设计师为保留日本文化中朴实无华的特点，使用天然毛面锈石及粗面花岗石等，又融入一些如镜子、马赛克等时尚材料。室内设计体现新旧交替、日西合璧之思路，较为时尚、自由，营造了一个多元化的饮食空间。入口的景观走道是从日本家喻户晓的民歌《北国之春》中得到灵感而采用斑驳无华的白桦木的。景观走道的设计不仅吸引着店内的顾客，对于店外的人们也能创造很强的视觉冲击。

As you enjoy the faint fragrance of white birch wood, automatic door slowly opens to welcome you, and then the atmosphere suddenly becomes warm. Multifunction hall employ more natural materials, wooden Japanese partition, wooden table and Japanese style chandeliers, giving people the feeling to recover one's original simplicity.

Box design uses mean manipulation. Soft colors, following the Japanese home furnishing decoration, warm atmosphere provides a comfortable home environment for the diners.

One of the highlights of the case is a hair stone passage to the toilet. The top mirror glass used enhances the space height, and at the same time, creates diverse spatial levels.

在感受"北国之春"那淡淡的白桦木余香的时候，餐厅的自动门也缓缓打开，气氛瞬间变得热烈起来。料理店的多功能大厅设计采用了更多的天然材料，木质的日式隔断、木质餐桌和具有日本特色的吊灯，给人以返璞归真的感觉。

包厢的设计采用了中庸手法，并不哗众取宠。色彩柔和，以日本家居装饰为蓝本，温馨的氛围为就餐者提供了舒适如家的环境。

通往卫生间的毛面锈石走道也是本案的一大亮点。顶上镜面玻璃的运用，视觉上提升了空间高度，同时营造出丰富的空间层次。

Sakata Japanese Restaurant Phase II

SAKATA酒田日本料理二期

Designer: Jiang Guoxing
Design Company: Xupin Design Decoration Engineering Co., Ltd.
Project Location: Kunshan, Jiangsu
Building Area: 380 m²
Main Materials: Ebony, Chinese black brick, woven glass, hair stone, stainless steel tube, etc.

设 计 师：蒋国兴
设计公司：苏州叙品设计装饰工程有限公司
项目地点：江苏昆山
建筑面积：380平方米
主要材料：黑檀木、中国黑砖、布纹玻璃、毛面锈石、不锈钢管等

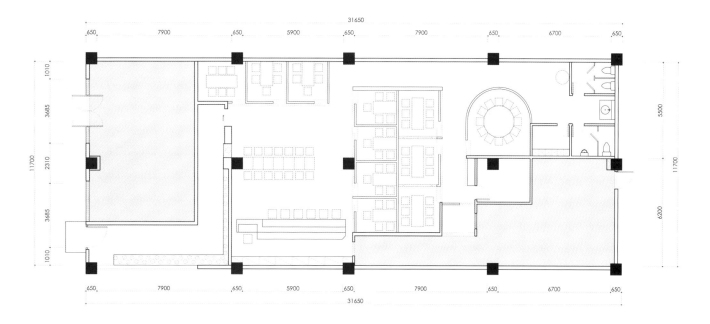

平面布置图

SAKATA Japanese Restaurant Phase II located in Hei Longjiang Road of Kunshan city, is an updated edition of phase I. Kunshan city is one of the strongest county-level cities in mainland China from the aspect of economy.

Designer of this case communicated with the client and renovated based on the structure of phase I. After having accumulated experiences for several years, the designer obtains more deep understanding about Japanese food culture. The designer interprets Japanese food culture in a modern and simple way, which keeps the content of Japanese culture in this modern and fashionable space.

SAKATA酒田日本料理二期位于昆山市黑龙江路，是在一期的基础上改造更新的产物。昆山是中国经济实力最强的县级市之一。随着经济的发展，日商大量涌入昆山，带动了餐饮业的发展，日本料理也倍受热捧，形成昆山主要餐饮类型之一。

本案设计师与业主沟通，在原一期结构上进行改造。经过几年的沉淀，设计师对日本料理文化有了更深层次的理解，运用现代手法与质朴的材质诠释日本料理文化，使得前卫时尚的空间内不失日本文化底蕴。

There's a view hallway at the entrance. On the surface of wall, the uneven raw wood bricks are used which extends to the end of hallway. The three-dimensional sense is added to the space with simple lines, giving powerful visual impact. The mottled white birches stand in lines on the other side of the hallway, making people calm and relaxed. At the ceiling mirror not only visually lifts the height of space, but also enriches spatial levels. What's more, the comparison and combination of the dynamic and static give customers a special sense. The elements of raw wood bricks have been extended. Natural rough stone, wooden Japanese partitions and dining tables let people feel being at home. And purple brown clothing tables and chairs together with black curtain, lessen the softness of the space, adding kind of romance to this simple fashionable pure room representing Japanese culture.

餐厅入口是一条景观走廊，墙面运用凹凸不平的原木木块饰面，延伸至走廊尽头，为简洁的线条空间增加立体感，并形成了视觉冲击。走廊的另一侧墙，斑驳朴实的白桦木林立成排，令人感觉沉静淡然，摒弃世俗杂事。顶部镜面的运用在视觉上提升了空间的高度，也使得空间更有层次感，动与静的碰撞结合给来客带来别样的感觉。在厅内，过道原木木块元素得到了延伸，质朴的毛锈石、木质的日式隔断以及餐桌给人返璞归真的感觉，而紫色咖啡色的布艺桌椅与黑色窗纱，调和了空间的柔和感，为这简约、时尚，带着朴实无华的日式文化气息的空间增加了一抹浪漫气息。

Traditional art of Japanese Tatami is used in VIP rooms. Materials need to be those with pure characters of Japanese culture. The ceilings feature wooden materials. Hanging ornaments of dead wood and branches express Buddhism gently and slightly. The whole colors are soft, together with dotted birch and flower. Designer does care about what customers need and let them feel at home.

It's fashionable, simple and pure, warm and cozy, gentle and elegant, and it serves exquisite Japanese food, the folk song of "The Spring of Northland", a unique Japanese atmosphere, all of which create this special Japanese food restaurant.

包间内选用日式榻榻米的传统工艺，材料运用上保留了日本文化朴实无华的特点，墙顶皆用木饰面，枯木干枝吊饰，让整个空间轻松淡雅，略带禅意。整体色彩柔和，加上桦木与花饰的点缀，让来客从中感受家的温暖。

这里时尚现代，这里简单质朴，这里温暖怡人，这里温柔雅致，这里有精致的日本料理，有《北园之春》的民间小调，有日式特有的文化气息，种种融汇成别样的日本料理餐饮空间。

Suzhou Sakata Japanese Cuisine

苏州酒田日本料理

Designer: Jiang Guoxing
Design Company: Xupin Design Decoration Engineering Co., Ltd.
Project Location: Suzou
Building Area: 450 ㎡

设 计 师：蒋国兴
设计公司：苏州叙品设计装饰工程有限公司
项目地点：苏州
建筑面积：450平方米

In a modern socialist technique, the designer uses new technology and materials to embody Japanese culture essence, together with refining language, rich subtle lighting change, simple and unadorned color contrast. Although it represents Japanese culture, yet it's filled with modern spirit and aesthetic appeal. It employs the materials freely, for example: a 3-meter-high sliding door made of steel, large areas of wall and ceiling; and the transluent black steel structure partition, forming a comparison between the virtual and real and kind of interest.

以现代主义手法，运用新技术、新材料表现了日式文化的本质特征。此外，运用了精练的语言，丰富微妙的光影变化，朴实无华的色彩对比。苏州酒田日本料理虽然取意日式文化，但充满了现代主义精神和审美情趣。对材料进行大胆地运用，比如：用钢构制作3米多高的移门，用大面积的复合地板铺设墙面及天花；通过半通透黑色钢构隔墙，形成虚实衬托，意趣盎然。

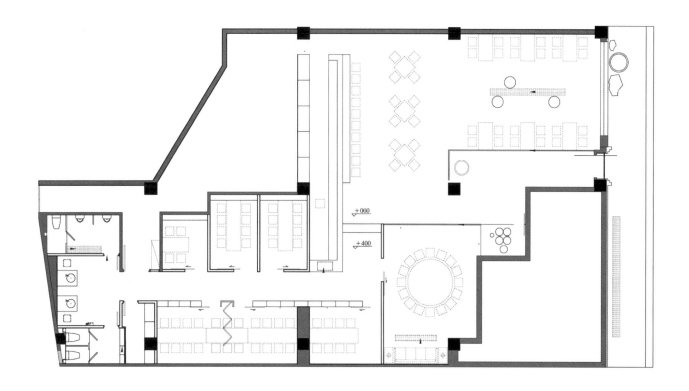

平面布置图

9.9inPizza Italian Restaurant

9.9inPizza意式餐厅

Designer: Jiang Guoxing
Design Company: Xupin Design Decoration Engineering Co., Ltd.
Project Location: Urumqi, Xinjiang
Building Area: 950 m²
Main Materials: Imported tiles, red wall brick, wooden flooring, solid wood batten, antique tiles, diatom mud

设 计 师：蒋国兴
设计公司：苏州叙品设计装饰工程有限公司
项目地点：新疆乌鲁木齐
建筑面积：950平方米
主要材料：进口花砖、红色通体砖、木地板、实木条、仿古砖、硅藻泥

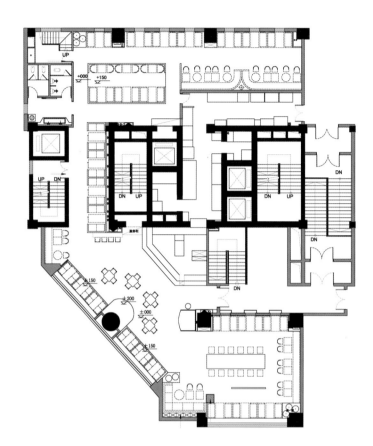

平面布置图

This case is an Italian restaurant of mixed style, the integration of the European countryside style and the modern minimalist style, adopting the simplest shape and the most primitive materials to create the most beautiful effect. For an Italian restaurant, the designer does not advocate the pursuit of the high-end luxury, but strives to achieve a unique feature, the elegant flavor and the casual comfort.

Distressed wooden flooring and red wall bricks are used in a large area, providing the restaurant with some nostalgia feelings in addition to its characteristics and fashion. White blinds and green partitions brighten the whole space and bring a fresh feeling, giving the customers a more comfortable dining environment.

It is a pleasant thing to enjoy the delicious Italian cuisine with friends in the leisure time.

　　本案为混搭风格,将欧式田园与现代简约融为一体。利用最简单的造型和最朴质的材料表达出最美的效果。对于一个意式餐厅,本案设计师并不主张追求高档豪华,力求个格独特、格调高雅而不失随兴舒适。

　　做旧处理的木地板与红色通体砖的大面积使用,使得整个餐厅在个性、时尚之余,多了几分怀旧的情怀。白色百叶窗与绿叶造型的隔断增添了整体空间的清新感觉,使人用餐时更为舒畅。

　　午后的休闲时光中,在这里享受着意式美食,与三两好友做伴畅谈,人生乐事不过如此。

Tangcheng Hot Pot City

汤城火锅店

Designer: Jiang Guoxing
Design Company: Xupin Design Decoration Engineering Co., Ltd.
Building Area: 1,100 ㎡
Main Materials: Wood flooring, gray lacquer, gray paint, gray wallpaper, black steel, black square tube
Photos: Jiang Guoxing

设 计 师：蒋国兴
设计公司：苏州叙品设计装饰工程有限公司
建筑面积：1 100平方米
主要材料：实木复合地板、灰色真石漆、灰色乳胶漆、灰色墙纸、黑色钢筋、黑色方管
项目摄影：蒋国兴

This case is a restaurant of mixed style, simple but luxurious, elegant but restrained. As a Chinese restaurant, its Chinese style decoration also reflects luxurious element. Therefore, this case becomes unique.

Dark colors domain the whole space, black leather sofa, black steel tube partition and dark-colored solid wood flooring make the whole design extremely stable. Meanwhile, the designers use diamond tiles and square tubes tactfully. Besides, the metallic European chandeliers and vividly colored oil paintings create bright and shining parts in the calm background. Black square tubes are used as partition, and edges of squares make the whole space more there-dimentional with clear and soft lines.

本案为混搭风格，简约奢华，雍容优雅而不失内敛。对于一个中式餐厅，中式风格装修又显出奢华的设计，让本案别具一格。
本案以深色系为主，黑色的皮质沙发、黑色的钢管隔断、深色的实木地板，都让整个设计显得极其沉稳。同时，本案设计师巧妙地利用地面的菱形拼贴和方管的交织，再加上欧式金属吊灯以及色彩鲜艳的风景油画，从形式上、色调上打造沉稳背景中的鲜亮局部。黑色方管作为空间隔断，以及格子的棱角让整个空间更有立体感，线条分明却又不失柔美。

平面布置图

Xupin Design Company Bar(Kunshan)

叙品设计公司酒吧（昆山）

Designer: Jiang Guoxing
Design Company: Xupin Design Decoration Engineering Co., Ltd.
Project Location: Kunshan, Jiangsu
Building Area: 80 m²
Main Materials: Brick-pattern wallpaper, dark wood flooring, mirror mosaics, brown textured marble, etc.

设 计 师：蒋国兴
设计公司：苏州叙品设计装饰工程有限公司
项目地点：江苏昆山
建筑面积：80平方米
主要材料：砖纹壁纸、深色木地板、镜面马赛克、啡网纹大理石等

Xupin Design Company Bar is a small bar for the staff of Xupin Design and transformed from the original accessory room. As a small bar of the design company, it bears a heavy responsibility and has a different meaning, providing the busy designers a relaxed leisure space.

The design style of this case is different from the common bars or the entertainment space. It adopts the Chinese style with the integration of modern elements and classical European elements to create a fashionable leisurely club with Chinese charm. The stylish mirror mosaic ceiling, the scholarly bookshelf, exotic paintings, the gramophone full of memories, and the fireplace perfectly blend with each other so that everything is so comfortable. You can taste the tea and wine here or chat with friends to recall the past or long for the future. What a joy of life!

So far, Xupin Design Company Bar attracts a lot of designers because of its reputation, and many activities, such as "Design Salon", "designer party", etc., are held here. We hope more and more designers will come to this small bar in their spare time, and we sincerely welcome their coming.

叙品设计公司酒吧是供叙品设计公司内部人员使用的小酒吧,是由原先的饰品库改造完成的。作为设计公司的小酒吧,责任重大,意义也不同,是设计师繁忙之余,放松心情的休闲空间。

本案酒吧的设计风格,不同于一般的酒吧娱乐空间,它以中式风格为主线,融入现代与欧式的经典元素,打造休闲时尚不失中式韵味的会所。时尚的镜面马赛克吊顶,充满书香气息的书架,异域风情的油画,充满回忆的留声机,给冬季带来温暖的壁炉等,这些元素在这里得到完美的融合,一切都是那么舒服。你可以在这品茶论酒,谈天说地,也可以与朋友回忆往事,畅谈未来。这些无不是人生的快事。

至今为止,叙品设计公司酒吧吸引了许多"慕名而至"的设计师朋友们,他们都愿意来这举办"设计沙龙"和"设计师聚会"等活动,叙品设计公司酒吧在业界已算是"小有名气"。希望越来越多的设计师朋友们能够在闲暇之余,步至叙品设计公司酒吧,我们在此诚挚地欢迎你们。

Xinjiang Huazhi Boiling Fish

新疆花枝沸腾鱼

Designer: Jiang Guoxing
Design Company: Xupin Design Decoration Engineering Co., Ltd.
Project Location: Urumqi, Xinjiang
Building Area: 850 ㎡
Main Materials: Wood veneer, wood lattice, litchi surface stone,
gray mirror, wooden flooring, diatom mud

设 计 师：蒋国兴
设计公司：苏州叙品设计装饰工程有限公司
项目地点：新疆乌鲁木齐
建筑面积：850平方米
主要材料：木饰面、木格、石材荔枝面处理、灰镜、木地板、硅藻泥

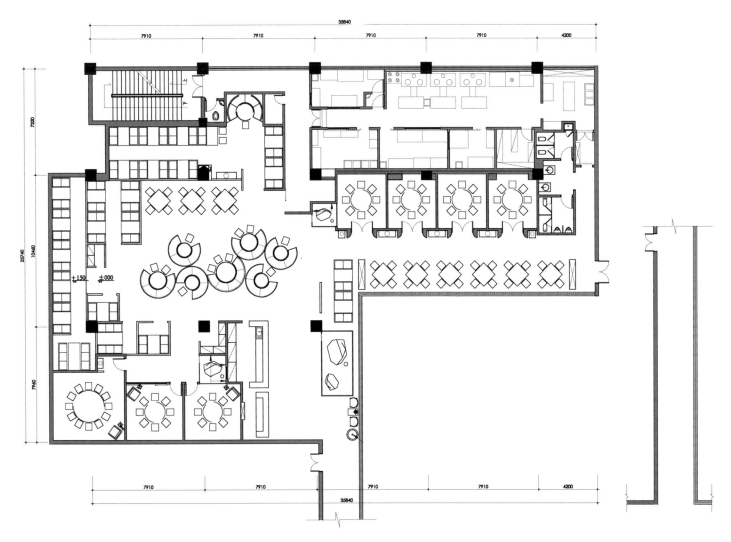

平面布置图

The case is located in Urumqi, with modern fashion elements and Oriental elements in the whole space, demonstrating a generous and profound oriental space.

The designer draws the inspiration from traditional oriental elements. The refined and elegant pots of blue-and-white porcelain are very charming and attractive, and the virtual-real contrast of the tactile diatom mud and the wooden partitions of Chinese style in the entrance achieves kind of relaxation. In the hall, the wooden partitions in the shape of birdcage are relatively independent and provide the privacy, yet full of unique characteristics. The aisle between the boxes is decorated with a row of rough yet generous golden doorframes of Chinese style, showing a royal style. Besides, the designer skillfully applies dense colors and line drawing paintings to make the space filled with good ventilation and the vitality. All these traditional oriental culture elements are never a simple display, but flamboyantly express the contemporary aesthetic through the contemporary design. In this imaginative space, the contemporary art encounters the traditional culture; the art collides with the space; the space is filled with vitality showing a romantic temperament. No doubt, this space is your psychological refuge zone.

　　本案位于乌鲁木齐，将现代时尚元素与东方元素深植于整个空间之中，蕴含大气、深邃的东方意境。

　　设计师从传统东方元素中汲取灵感，使用精致典雅的青花瓷盆，意趣盎然。入口处富有质感的硅藻泥与中式木隔断虚实对比。大厅一座座鸟笼形木条隔断相对独立私密而又独具个性；包间走道里一排粗犷大气的中式金色门框，颇有皇家气派。设计师炉火纯青地运用厚重色彩，加以配饰如白描挂画使得空间生机勃勃。这些传统东方文化元素绝不简单罗列，而是通过当代设计形式张扬地表达当下的审美情趣。在这个充满想象的空间里，当代艺术和传统文化邂逅，艺术与空间碰撞，生命在空间里盈动，而拥有一种浪漫的气息，成为你心灵皈依的地带。

Refined Food Restaurant

精膳

Designer: Jiang Guoxing
Design Company: Xupin Design Decoration Engineering Co., Ltd.
Project Location: Urumqi, Xinjiang
Building Area: 1,700 ㎡
Main Materials: Corrugated gray stone, black and white marble, dark solid wood flooring, dark gray brick, oak wood veneer, maroon linen, etc.

设 计 师：蒋国兴
设计公司：苏州叙品设计装饰工程有限公司
项目地点：新疆乌鲁木齐
建筑面积：1 700平方米
主要材料：灰色波纹石、黑白根大理石、深色实木地板、深灰色砖、橡木木饰面、枣红色亚麻布等

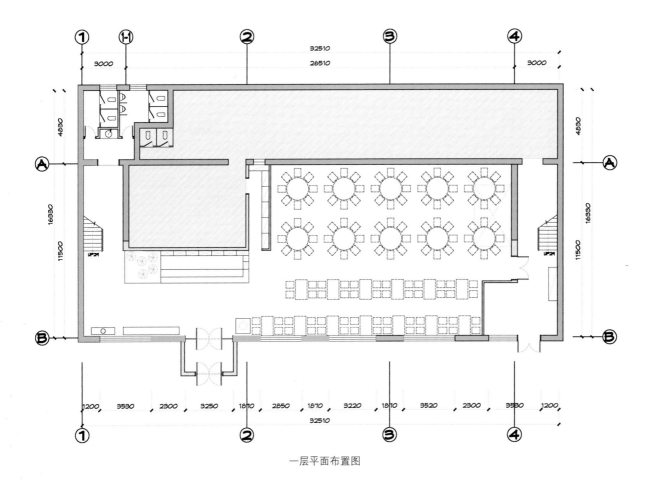

一层平面布置图

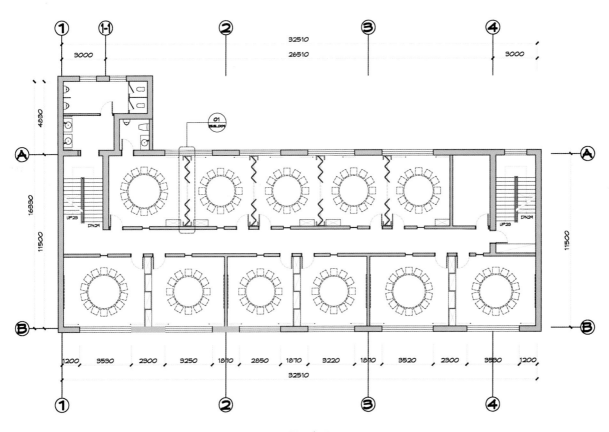

二层平面布置图

Refined Food Restaurant is located in Tianjing Road in Urumchi city of the Xinjiang Uygur Autonomous Region, serves food of Xiang style (of Hunan province) and has an excellent location. Designers combine traditional Chinese style and modern elements, and the leisure atmosphere dominates the space which is not limited by Chinese style. Grey and maroon are main colors of the restaurant, complemented by black. The furniture which is a little bit nostalgic, together with soft light, creates a warm and cozy place.

精膳位于新疆乌鲁木齐天津路，地理位置得天独厚，主营湘菜。设计师将传统中式元素和现代元素相结合，以休闲氛围为主调，令空间不局限于中式。色彩方面，设计师以灰色、枣红色为餐厅的主要色调，黑色做辅色，加上略带怀旧风格的家具，再配上柔和的灯光，营造出十分温馨舒适的氛围。

Corrugated grey stones are used to bar counter at the entrance, which brings the dynamic effect. The ceiling is white and naked, which keeps the original height of space. While the walls are decorated with grey wallpaper and maroon linen, showing the passionate culture from Hunan province, which makes guests feel being at home. The left side of the entrance is the passage to upstairs. The corrugated stones cover the whole wall, which is solid and let guests not be bored when going upstairs. VIP rooms are on the 2nd floor. There're views of dried wood and stones with artistic shapes. The mirror at the back also extends the space of passage when reflecting the views. After stepping into VIP room, you can feel very quiet, which is different from noisy hall. VIP rooms adopt grey wallpaper and simple lines, together with clay pot and ink paintings on the wall. Simplified and fashionable modern Chinese chandeliers not only promote the aesthetics, but also make the room vivid, enriching the visual effect.

Nowadays, people are paying more and more attention to the cultural atmosphere and individuation of dinning space when enjoying delicious cuisine. And having food in this simplified, bright colored, quiet VIP room can meet the requirement of guests. Designers use color as a language to show us the simplicity and elegancy of culture of Hunan province, presenting the humane and unique design.

入口吧台运用灰色波纹石使得吧台更具动感，大厅顶棚做成白色裸顶，保持了空间的高度。墙面则用灰色壁纸，枣红色亚麻布运用在大厅中彰显湖湘文化的热情，整个大厅充满着浓浓的暖意，使宴会客人在热闹的情景下更感受到家的温暖。在入口左侧是餐厅的上楼通道，设计师用波浪石铺满侧面墙，以增强立体感，这样客人上楼时就不会感觉沉闷。二楼是餐厅大包间区，包间过道的尽头，布置了景观，枯木石景，用景写意，背面的镜子在反射景观的同时也在视觉上延伸了过道的空间。步入包间，让你感觉到的是宁静的氛围，有别于大厅的火红热情。包间运用灰色壁纸与简练的线条，加上陶罐装饰、墙壁中式泼墨挂画宁静致远，而简约时尚的现代中式吊灯又使得包间更加活跃，丰富了包间的视觉效果。

现代人在品味美味佳肴的时候，开始关注用餐环境的文化氛围和个性化。这种简洁、色彩热烈、安静的包厢环境，恰好满足了顾客这部分的需求。设计师用色彩体现湖湘文化背景，并运用中式文化的简练和韵味，使其更具有人文气息。这是本案设计师独具一格的设计。

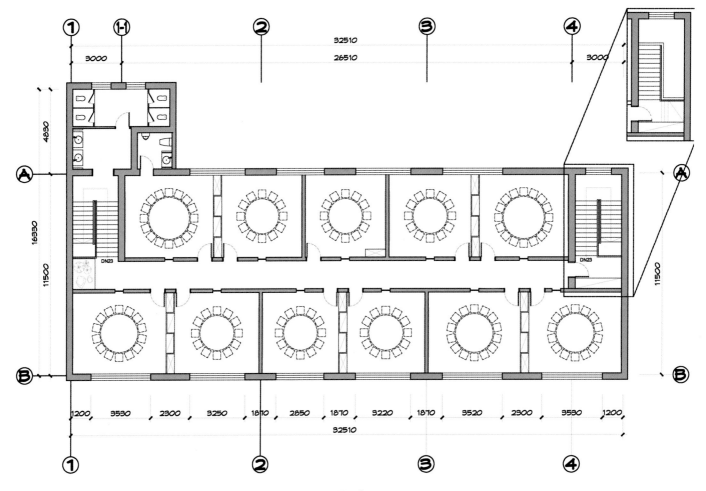

三层平面布置图

Menyingtianxia Hot-Pot Restaurant

门迎天下火锅店

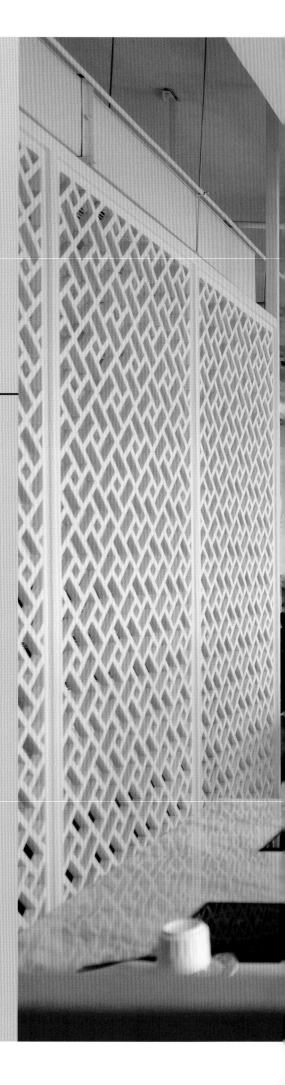

Designer: Jiang Guoxing
Design Company: Xupin Design Decoration Engineering Co., Ltd.
Project Location: Urumqi, Xinjiang
Building Area: 900 m²
Main Materials: White texture wallpaper, white wood Flooring, microcrystal brick, Chinese wooden lattice, imitation knight white marble tile, mirror strip, etc.

设 计 师：蒋国兴
设计公司：苏州叙品设计装饰工程有限公司
项目地点：新疆乌鲁木齐
建筑面积：900平方米
主要材料：白色肌理壁纸、白色木地板、白色微晶砖、中式木格、仿爵士白大理石砖、条镜等

Menyingtianxia is an upscale hot-pot restaurant located in a flourishing commercial street of Urumuqi City, the Xinjiang Uygur Autonomous Region. Hot-pot is one of the most popular catering categories and benefits from the advantaged geographical environment there. The designer makes full communication with the owner about how to make the restaurant stand out from the industry. The hot-pot restaurant themed on "Blue and White Porcelain" appeared in response to the thought of creating special food products and unique space environment.

The designer uses the color of the blue and white porcelain to make a bright, decent and noble dining space. Blue and white porcelain, is one of the mainstream space decoration in the ancient China, and it also plays an important part in modern collection of cultural relics. Most of the blue and white porcelain feature white background and blue stripes. They are clear, bright, lovely and beautiful.

"门迎天下"是一家中高档火锅店,位于新疆乌鲁木齐一条繁华的商业街上,火锅是新疆首府主流的餐饮类型之一。设计师与业主沟通,怎样在餐饮行业中脱颖而出,特立独行。以打造特色的餐品与独特的空间环境的思想为指导,这个以"青花瓷"为主题的火锅店应运而生。

本案设计师以"青花瓷"的色彩为主题,意境作景,打造一个明朗、大方、高贵的餐饮空间;"青花瓷"又称"白地青花瓷",简称"青花",是中国古代主流的空间饰品及现代收藏文物之一,多为白底蓝纹,蓝白相映,晶莹明快,怡然成趣,美观隽久。

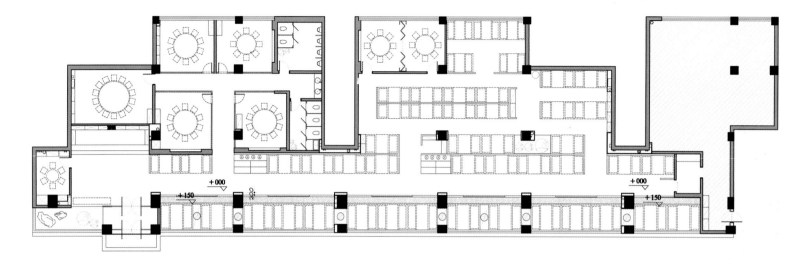

平面布置图

The main color White and the complementary color blue together with the Chinese space elements create a new Chinese style, which is the innovation of the designer. The whole layout is clear and smooth. It is the reception desk and the hall that come before you after entering the entrance. The roof of the hall is decorated with mirror strips, which create an artistic effect and enlarges the space at the same time. The white scattered seats make people feel cheerful. The white roof, the white flooring, the white tables together with the blue birdcage-shaped lamps and the blue window frame display nobility and elegance. Private rooms are on the both sides of the white lattice passage, which also reflect the blue and white porcelain style. The silver roof, the white texture wallpaper and the blue and white decorations reflect the nobility and elegance perfectly.

The beauty of modern Chinese style together with the blue and white seems to tell a beautiful story in the distant past. All of these create a quiet, beautiful, bright and elegant dining space, which will make customers reluctant to leave.

整个空间中白色是主体色，蓝色为辅色，融入中式空间的空间元素，衍生出一种新的中式风格，这对于设计师也是一种创新。餐厅的整体布局动线清晰、流畅。入口进门是前台与大厅，大厅顶部以条镜修饰，既美观，又拉伸了空间高度，在右尽是清爽开敞的白色散座区，顿时让人有豁然开朗之感，白色的裸顶，白色的地砖，白色的餐桌相间蓝色的鸟笼灯，蓝色的窗框饰面，无不流露出高贵素雅之感。沿着精致的半通透白色木格过道，是餐厅的包间区，包间里延续了大厅的蓝白色系，银箔饰顶，白色肌理壁纸，加之青花饰品的点缀，高贵素雅的气质呼之欲出，与大厅遥相呼应。

蓝色的意，白色的境，加之现代中式的美，仿佛诉说着遥远过去的一段美丽故事，形成了一个幽静美观、明净素雅的餐饮空间，令人流连忘返。

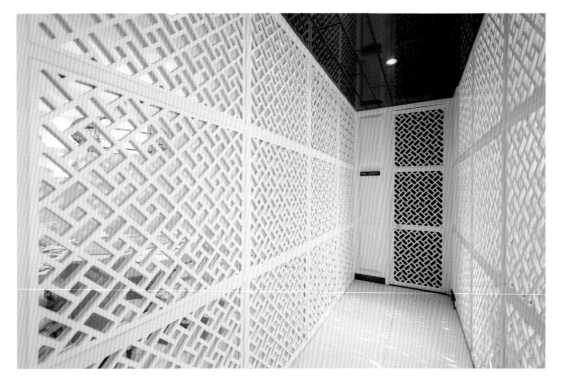

Bamboo Stream No. 1 Restaurant

竹溪一号

Designer: Jiang Guoxing
Design Company: Xupin Design Decoration Engineering Co., Ltd.
Project Location: Urumqi, Xinjiang
Building Area: 1,200 ㎡
Main Materials: Gray antique bricks, white diatom mud, dark wooded flooring, light-colored floor tiles, blue antique brick, oak wood veneer, dark green linen, etc.

设 计 师：蒋国兴
设计公司：苏州叙品设计装饰工程有限公司
项目地点：新疆乌鲁木齐
建筑面积：1 200平方米
主要材料：灰色仿古砖、白色硅藻泥、深色实木地板、浅色地砖、蓝色仿古砖、橡木木饰面、墨绿色亚麻布等

Bamboo Stream No. 1 Restaurant mainly serves Chinese Food and is located in Nanhu Road of Urumqi City in the Xinjiang Uygur Autonomous Region. This road is full of people and boasts an excellent position.

The highlight in this restaurant is the color of bamboo-dark green, reflecting the name itself. Bamboo is long and straight, always green in winter. Bamboo is one of the favorite plants of the Chinese. Considering all above reasons, designers adopt dark green of bamboo as main color in the whole space, which makes the restaurant a refreshing, vibrant, happy, fresh and cozy atmosphere.

竹溪一号餐厅，位于新疆乌鲁木齐南湖路，地理位置得天独厚，人气很旺，是一家主营中餐的特色餐厅。

竹溪一号餐厅最大的特色就是竹色——墨绿色，正好又能关联到"竹溪一号"中的"竹"这个字。竹子因其很长、笔直且冬季依旧翠绿，素有四君子（梅兰竹菊）之一和岁寒三友（梅松竹）之一的美称，备受中国人喜爱。正因为这些，设计师在整个空间都以竹色为主调，在整个餐厅营造出一种清新、充满活力、快乐、新鲜、舒适的氛围。

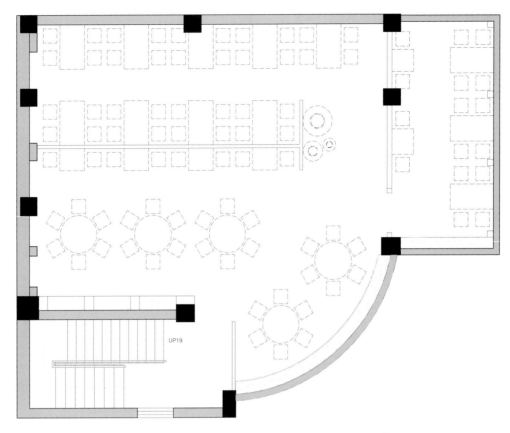

负一层平面布置图

Designers combine traditional pastoral style with modern elements to create a fresh, happy atmosphere, which let the space be smart. Dark green is the main color of the restaurant, while white and yellow are also adopted to complemeat it. Slightly retro furniture matches soft light to make a cheerful and fresh space.

Big insulation windows are built to separate the space of corridors and VIP rooms. There're also folding screens dividing space in big VIP rooms, which not only makes good use of space, and also extends and renovates the method of insulation windows. The combination of dark green artificial beam and white painted oak wooden battens makes the space more interesting and beautiful. Natural materials like dried wood, pebbles, bamboo and webbings are employed to decorate the hall and corridors. Display of green bonsai, porcelain and pottery are with the shape and spirit of traditional decoration, creating cultural taste and image, also showing unique charm of Chinese traditional cuisine.

"Back to nature" is the slogan in pastoral style of modern restaurant. In fact, only combining with the law of nature can we get balance between the physical and mental in current society with so fast pace. You can relax yourself and enjoy the dining environment. Designers present all these elements incisively and vividly in this case.

本案中设计师将传统的田园风格和现代的元素相结合，营造清新、欢快的氛围，令空间不呆板。色彩方面，设计师以墨绿色为餐厅的主色，白色、黄色为辅色，加之略带复古的家具，再配上柔和的灯光，使整个空间十分欢快、清新。

其中过道、包间多采用大的隔窗来分割空间，并且大包间采用屏风隔断划分空间，不仅使得空间得到合理的利用，而且使得隔窗这种设计手法得到延续和创新。天花采用墨绿色的假梁和白色橡木木拼条的结合，以提升空间的趣味和美感。大厅、过道在软装饰上均使用枯木、鹅卵石、竹、织物等天然材料装饰；绿色盆栽、瓷器、大陶罐等摆设，吸取传统装饰"形"、"神"的特征，让餐厅更具有文化韵味和意境，体现出中国传统餐饮文化的独特魅力。

现代餐饮中的田园风格设计倡导"回归自然"，结合自然规律，才能让客人在当今快节奏的社会生活中获取身体上和精神上的平衡，就餐时放松心情，以愉悦的心情享受就餐环境。

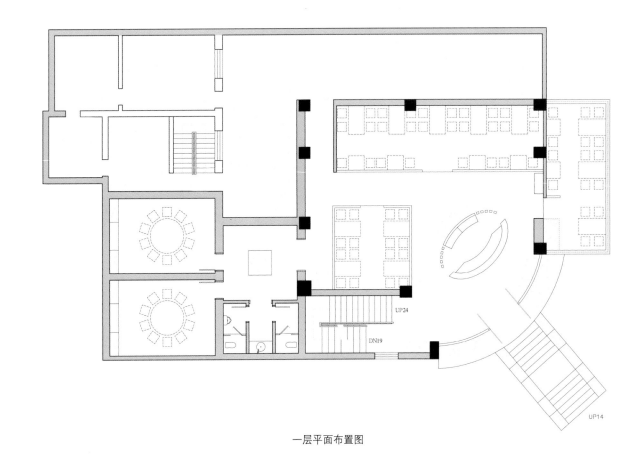

一层平面布置图

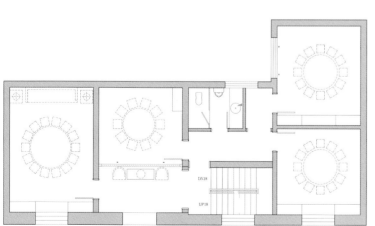

二层平面布置图

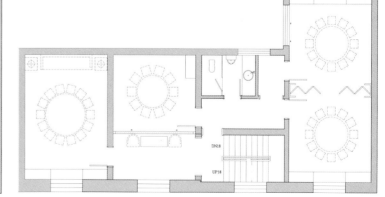

三层平面布置图

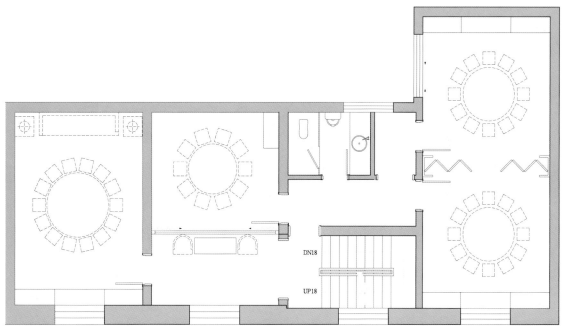

四层平面布置图

Mashijiu Pot Restaurant on Danlu Road

马仕玖煲丹露店

Designer: Jiang Guoxing
Design Company: Xupin Design Decoration Engineering Co., Ltd.
Project Location: Urumqi, Xinjiang
Building Area: 700 ㎡
Main Materials: Beige microlite, white square tube, tawny mirror, line drawing paintings, gypsum board, dark wooden flooring, birdcage lamp, horse-head-lantern, etc.
Photos: Jiang Guoxing

设 计 师：蒋国兴
设 计 公 司：苏州叙品设计装饰工程有限公司
项目地点：新疆乌鲁木齐
建筑面积：700平方米
主要材料：米白微晶石、白色方管、茶镜、白描挂画、石膏板、
　　　　　深色木地板、鸟笼灯、马头灯等
项目摄影：蒋国兴

It is a chain brand of Chinese food in Xinjiang, and the case is located in the commercial plaza in the downtown of Urumqi. The target consumers of the restaurant are mainly white collars, so the design highlights the modern business catering and integrates with the surrounding high-end international brands.

In the case, designers employ the modern technique, white and gray, and a rational and generous combination of the space and elements. The succinct strokes, pure approaches, and ordinary materials create an impressive spatial effect, express a connotative and unique temperament. The new interpretation of Chinese style is matched with the unified color, unique birdcage lamps, white chairs of Chinese style and other elements to create a poetic mood and an exquisite modern urban lifestyle.

　　马仕玖煲是新疆的一个中餐连锁品牌，本案位于乌鲁木齐市中心商业广场。该店消费对象主要为时尚白领，本案设计重点突出现代商务餐饮与周边高端国际品牌的呼应与融合。

　　本案采用简洁的现代的手法，运用白色、灰色，将空间和元素进行合理、大气的组合；以洗练的笔触，纯净的手法，用普通的材料营造动人的空间效果，表现出极富内涵和独特的空间。对中式的风格进行了全新的诠释，用整体统一的色彩搭配，造型独特的鸟笼灯，白色中式椅等元素营造出诗情画意，体现现代都市精致的生活品位。

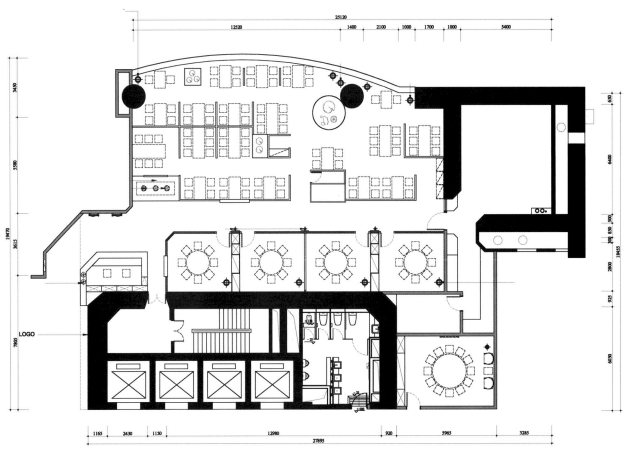

平面布置图

The latticed gray mirror wall of large areas gives the limited space a concise and generous sense. The ingenious layout of a half white Chinese chair on the tawny mirror wall attracts people to ponder yet is full of joy. Besides, the scene of pebbles and horse-head-lanterns is also unforgettable. A neat row of fashionable birdcage lamps in the aisle make the entire space well-stacked, demonstrating a balanced virtual-real situation. The partitions of white square tubes create a private and relaxed dining space.

本案大面积采用木格灰镜墙面，让不大的空间显得简约大气。设计师别出心裁地把半张白色中式椅挂在茶镜墙面上，充满趣味。恰到好处的鹅卵石、马头灯造景令人难以忘怀。过道一排整齐的时尚鸟笼灯让整个空间饱满充实，虚实相称。白色方管隔断，营造一种既私密又空旷放松的就餐氛围。

Yuanshan I

原膳一期

Designer: Jiang Guoxing
Design Company: Xupin Design Decoration Engineering Co., Ltd.
Project Location: ulumuqi, Xinjiang
Building Area: 5,000 ㎡
Main Materials: Black granite (litchi surface processing), wood veneer, cultural stone, light gray wallpaper, dark wood flooring, metal screens, etc.
Photos: Jiang Guoxing

设 计 师：蒋国兴
设计公司：苏州叙品设计装饰工程有限公司
项目地点：新疆乌鲁木齐
建筑面积：5000平方米
主要材料：黑色花岗岩（荔枝面处理）、木饰面、文化石、浅灰色壁纸、深色木地板、金属帘等
项目摄影：蒋国兴

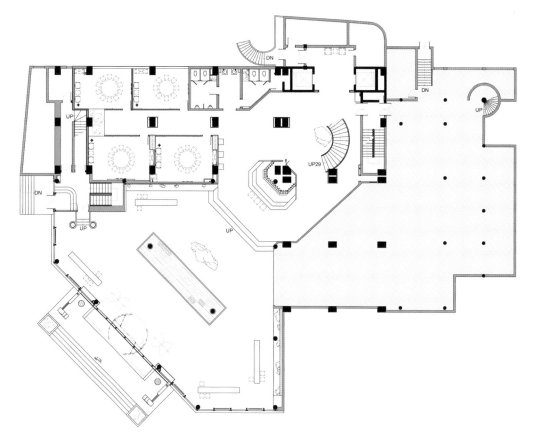

一层平面布置图

In this case, the designer adopts the dark tone to create the Oriental Zen mood and the spatial layers with simple but tough lines. The techniques of the virtual-real alternation, the change of light and dark, and the dialectical combination of simplicity and complexity bring a refreshing charm of ancient rhyme.

The designer does not rigidly adhere to the demonstration of the details, but skillfully uses some Chinese elements to heighten the mood, giving people more space to ponder and taste. The artistic furnishings of the Oriental theme can be as elegant as a gentleman and also as noble as an emperor, which can be understated or highlighted, full of flavors.

The background wall in the entrance with the creative white concentric circles gives a strong visual impact, which shows a free and easy style of changing the Earth's gravity. The well-proportioned scene of water and birdcages, elegant and quaint lamps made from traditional Xuan paper, ceramic lacquer products, and quiet drawing paintings outline a scene which is faintly discernible at first glance yet unforgettable. The boxes of large areas are decorated with simple and elegant modern furniture of Ming Dynasty style, the landscape with profound meanings made from dead wood, and the black leather sofa revealing calm luxury, so the boxes are simple, quiet and intoxicating. In conclusion, the designer uses the Zen charm to present the interior design, aiming to show the communications between the human and the nature and create a soul habitat for modern people.

本案设计师采用深色基调营造东方禅韵，用简洁硬朗的直线条勾勒出富有层次感的空间；以朴实的手法，通过虚与实，明与暗，简与繁的辩证结合实现一种"古韵新风"。

设计师并不拘泥于细节的刻画，恰到好处地运用一些中式元素来烘托意境，给人更多的留白空间去品味思索。东方主题的艺术陈设可以淡雅如君子，可以贵气如帝王，可以轻描淡写，可以浓妆重抹，各领风骚。

入口处极具创意的白色同心圆背景墙从视觉上给人强烈的冲击，有种改变地球引力的洒脱。错落有致的流水鸟笼造景，大气古朴的传统宣纸灯具，让人兴趣盎然的陶瓷漆器制品，宁静致远的白描挂画，这些元素勾勒的景象，乍看若有若无，却让人难以忘怀。超大面积的包厢，配以简洁大方的现代明式家具，意味深远的山水枯木造景，奢华的黑色皮质沙发。整个包厢简约宁静，让人陶醉，以禅韵来诠释室内设计，旨在体现人与自然的沟通，以求为现代人营造一片灵魂的栖息地。

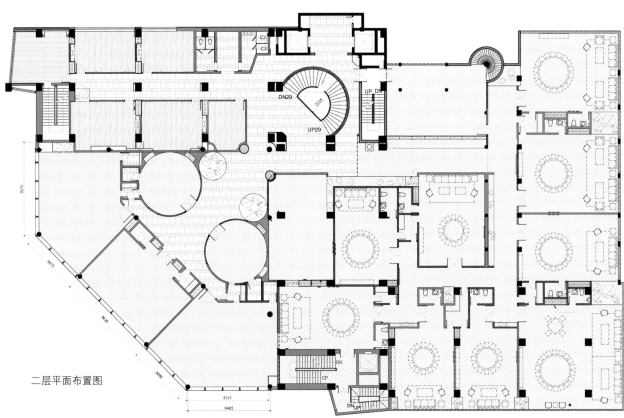

二层平面布置图

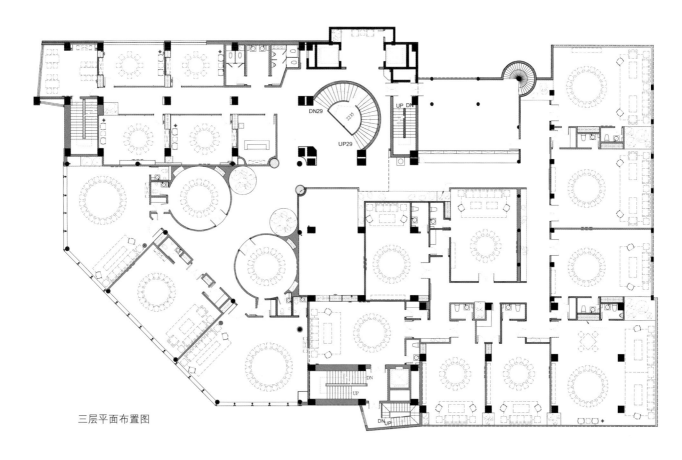

三层平面布置图

Yuanshan II

原膳二期

Designer: Jiang Guoxing
Design Company: Xupin Design Decoration Engineering Co., Ltd.
Building Area: 1,250 ㎡
Main Materials: White travertine, wood veneer, beige wall coverings, dark wood flooring, metallic screens, etc.
Photos: Jiang Guoxing

设 计 师：蒋国兴
设计公司：苏州叙品设计装饰工程有限公司
建筑面积：1 250平方米
主要材料：白洞石、木饰面、米白色壁布、深色木地板、金属帘等
项目摄影：蒋国兴

Yuanshan II located in Urumqi, Xinjiang is a continuation of Yuanshan I, so it keeps the refreshing charm of ancient rhyme in the Yuanshan I, but uses a different way to express the Oriental Zen mood. The new Chinese style in the case does not purely pile up elements, but combines modern elements with traditional elements through the designer's understanding of the traditional culture. The designer decorates the space of the traditional flavor with the aesthetic needs of modern people so that the traditional art gets a right expression in today's society.

The entire aisle is designed with a dark tone. The designer deliberately only takes the lighting from the floor candelabra in an orderly layout to invisibly create an Oriental mystery. Walking through the aisle, you will encounter the ornamental landscape of water, hill and dead wood which uses the natural materials to interpret a rationality of Zen, bringing the quietness to your mind. The boxes overturn the application of the dark tone in the former space, by taking beige-white as the main color. The gently elegant beige-white wall coverings, tough straight lines, restrained and simple Chinese furniture, Chinese paintings with profound meanings, and quaint books give people a feeling of warmness and elegance and also a graceful sense. Through the windows of the boxes, you can catch a piece of bamboo landscape, and the bamboos are tall and green, elegant and noble, graceful and perseverant. What a magnificent natural beauty! In such a space, you can feel the natural serenity so as to help you get rid of the tediousness.

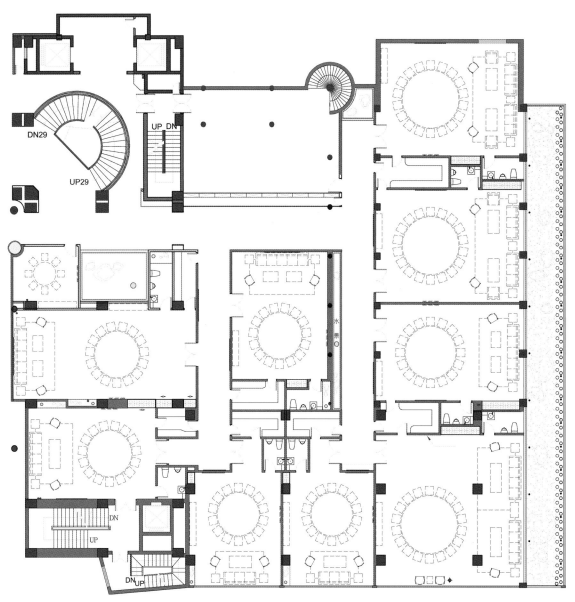

平面布置图

　　位于新疆乌鲁木齐的原膳二期是原膳一期的延续，它保留了一期的古韵新风，却以不同手法表达了东方禅韵。本案的新中式风格不是纯粹的元素堆砌，而是通过对传统文化的认识，将现代元素和传统元素结合在一起，以现代人的审美需求来打造富有传统韵味的事物，让传统艺术在当今社会得到合适的体现。

　　本案设计师运用深色基调打造整个过道，在灯光方面故意只靠排列有序的落地烛台取光，无形中营造了一种东方神秘感。徒步过道之上的你还会遇到山水枯木造景，它选用天然的装饰材料来诠释禅宗的理性，带给你宁静安逸。包间颠覆了以往由深色基调勾勒的空间，运用了以米白色为主色、温柔素雅的壁布，简洁硬朗的直线条，质朴的中式家具，意味深远的中式画，古香古色的书籍，这些都使人觉得温馨淡雅而不失大气。包间橱窗外可以看到一片竹林造景，它们挺拔苍翠、坚忍不拔、典雅高洁、婀娜多姿，形成了一幅壮美的诗意画卷。在这里你可以感受到自然与宁静，脱离城市的烦琐。

Shenhai Yihao

深海壹號

Designer: Jiang Guoxing
Design Company: Xupin Design Decoration Engineering Co., Ltd.
Project Location: Urumqi, Xinjiang
Building Area: 2,000 ㎡
Main Materials: Dark-colored solid wood flooring, wood veneer, black granite, grey granite, dark blue tiles, blue wallpaper, gold foil, etc.

设 计 师：蒋国兴
设计公司：叙品设计装饰工程有限公司
项目地点：新疆乌鲁木齐
建筑面积：2 000平方米
主要材料：深色实木地板、深色木饰面、黑色花岗岩、灰色花岗岩、深蓝色砖、蓝色壁纸、金箔纸等

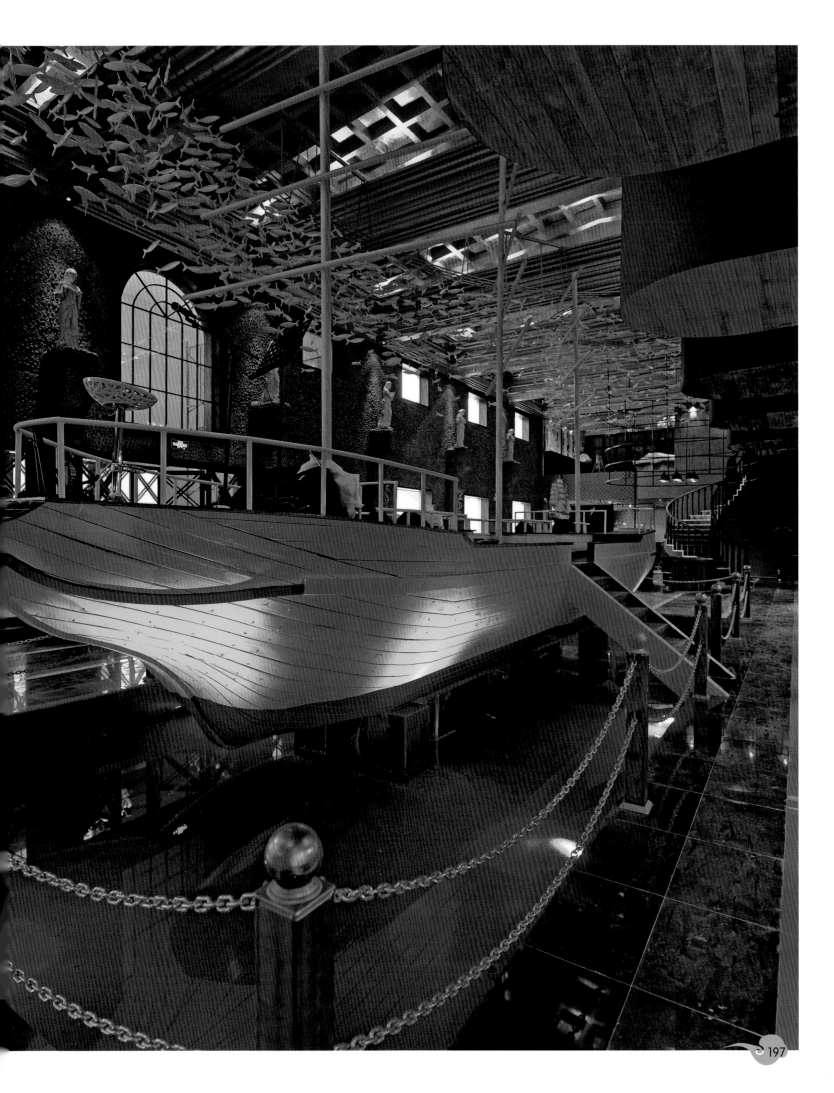

Shenhai Yihao is located in the Bar Garden, South Lake Road, Urumqi, Xinjiang. This is a new commercial district integrating entertainment with dining. It is a high-end restaurant themed on ocean and specialized in seafood hotpot.

深海壹號位于新疆乌鲁木齐南湖东路酒吧园区,该园区为近年兴起的商业区,集娱乐、餐饮为一体。该店主营海鲜火锅,是一家以海洋为主题的高端餐厅。

After communications with the owner, the designers have a deeper understanding of the ocean. They want to create an ocean-themed restaurant by conception skills.

As we all know, different themes of culinary culture offer different feelings, ocean-themed restaurants offer us a blue feeling. The blue usually associates to broadness, deepness, and the blue sky, waves, which altogether make us feel elegant and serene. Gold is the second tone of the restaurant, which is mainly located on the roof. At the background of the original structure, the roof is decorated by decorative strips and warm light to create a visual effect of brightness, magnificence and glory. The gold is like tender beach. The beach and the ocean complement each other. The waves come and go quietly. Listening to the breeze, and feeling the tenderness of the sand, one can enjoy the wonderful time.

业主与设计师沟通，对海洋概念的了解更深，形成一个用概念手法诠释海洋文化的主题餐厅。

不同的餐饮文化主题给人不同的感受，以海洋为主题的餐厅蕴含着一种蓝色的氛围。蓝色调体现着丰富的情感内涵，蓝色让人联想到广阔深远的天空，波涛滚滚的大海。忧郁的情感总是与蓝色连在一起，使人感觉优雅、宁静。金色是餐厅的第二色调，主要分布在餐厅顶部，顶部保留原结构造型，增加线条装饰，配以暖色灯光，拥有产生光明、华丽、辉煌的视觉效果。金色宛如轻柔的沙滩，沙滩与海相辅相成，海浪静悄悄地通过，又悄悄地退去，客人在此聆听着海风，感受着柔软的沙滩和温暖的太阳，享受着美好时光。

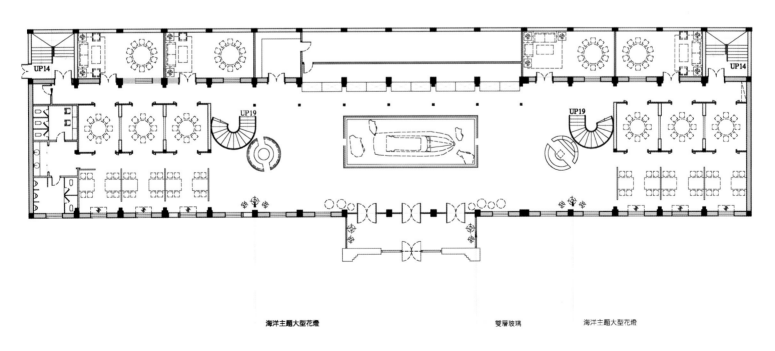

海洋主题大型花灯　　　　　　　　　双层玻璃　　海洋主题大型花灯

一层平面布置图

At the entrance of the hall, you can see a distressed ancient sailing boat. At the prow, there is a mermaid sitting on the rock. The floor is made of colorful and stripe-shaped marbles to increase the sense of depth. As if we are crossing boundless ocean, when walking on it. There are lots of lifelike silver fish over the large-sized boat. The wave, beach, fish, ship, whistle, deep-ocean-organism-shaped paper lanterns and the old captain all remind us of childhood. On the right is the service area which is decorated with incandescent bulbs and blue light. The wonderful colors create a fantastic atmosphere. On the left is a waiting area, you can take a seat on the round sofa, chat with your friends and appreciate the programs arranged by the owner. The casual sitting area and open rooms are on two sides, the scaly separated seats and the blue linen sofa, are full of "ocean" flavor. Walls of the open rooms are decorated with light bulbs which seem like the foams, floating on the ocean. The walls coupled with blue light create a romantic atmosphere. Blue and white chairs make the room active. The selected tableware also contributes to a noble and elegant dining environment. At the background of the original structure, the roof is decorated by gold foil and gypsum strips. The blue wallpaper and the sea creature decorations make the quiet space active. The private rooms are not totally-enclosed. The dining process will be more enjoyable. The surrounding environment will be filled with cultural flavor and intellectual atmosphere, which will satisfy the customers' tastes and expectations, and offer them an elegant and comfortable sense.

In tasting the delicious food, modern people also begin to pay attention to the cultural atmosphere and personalization of dining environment. The design is to meet customers' demand. The charming restaurant is quiet, mysterious, and fantastic. Through seeing, hearing and imagining, one can enter an expected paradise created by the design.

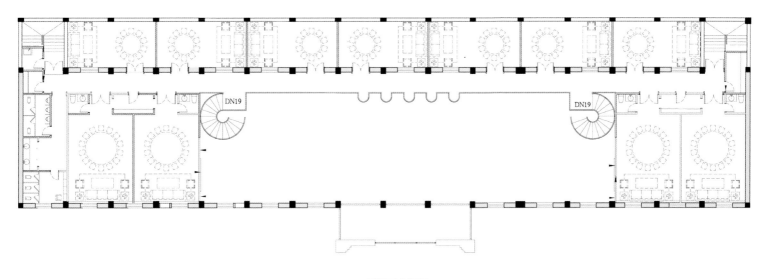

二层平面布置图

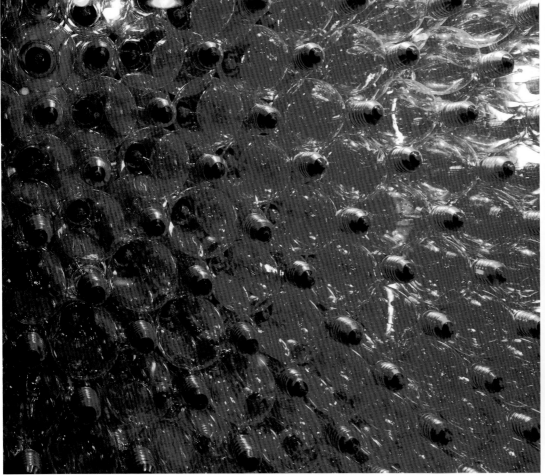

走进大厅入口，映入眼帘的是一只做旧的大型古帆船，经过帆船你会观赏到船头礁石上美人鱼的动人姿态，布局上大厅动线流畅，没有阻挡，而地面则采用各色条形大理石铺设，增加了大厅的层次感，犹如海浪，让人觉得自己不是在走路，而是驾着一艘舰艇穿行在辽阔无际的海洋里，自由而又惬意。帆船的上方，悬挂着成群的银鳞小鱼，如有生命的鱼儿嬉戏追逐，成群结队，让人更加亲近自然、热爱生活。海浪、沙滩、鱼群、古船、鸣笛、深海生物造型的纸灯还有老船长，这些都勾起你对童年的回忆。大厅右侧是服务台区，服务台前部由白炽灯泡装饰，配上蓝色灯光，流光溢彩，有种梦幻感。左侧设置了一个等候区，圆形的卡座沙发融入环境，休闲舒适，来客可以在餐前在这里小憩一会儿，可以与友人聊聊天，也可以欣赏餐厅主人为你安排的音乐节目。再往两侧去是餐厅的散座区和开放散包，散座以鱼鳞状隔断分隔，蓝色的麻布沙发，充满"海洋"的气息。而开放散包用灯泡装饰隔墙，宛如深海的泡沫向上漂浮，配上蓝色灯光营造的氛围，浪漫迷人。蓝白的椅子增加此空间的活跃感，再加上精心挑选的餐具，这些共同营造出高档雅致的用餐环境。包厢部分集中在餐厅二楼，大包厢的顶部延续了大厅的元素，保留原结构以金箔贴顶并加以石膏线条，蓝色的壁纸墙面，凸显海洋气息，并将活泼好动的海洋生物饰品融入环境当中，为安静的空间增添了趣味性。包厢并不是全封闭的，在包厢靠大厅的侧面墙，设计师运用鱼鳞隔断预留位置分隔，让来客在用餐的同时也可以欣赏大厅的音乐节目，增加用餐乐趣。将周围的环境应用与调动起来，充满人文气息的知性风格，融入了人们对生活的品味和期许，优雅舒适。

现代人们在品味美味佳肴的时候，开始关注用餐环境的文化氛围和个性化，而餐厅的设计正满足了顾客这部分的需求。这个像大海一样不可测度的魅力餐厅，宁静、神秘而新奇，希望在你身临其中的时候，通过视听与联想，能进入主题情境。

Mashijiu Pot-Aksu Restaurant

马仕玖煲阿克苏店

Designer: Jiang Guoxing
Design Company: Xupin Design Decoration Engineering Co., Ltd.
Project Location: Urumqi, Xinjiang
Building Area: 1,800 ㎡
Main Materials: Black granite, square steel, gray brick, black brick, gray wallpaper, blue fabric, etc.

设 计 师：蒋国兴
设计公司：苏州叙品设计装饰工程有限公司
项目地点：新疆乌鲁木齐
建筑面积：1 800平方米
主要材料：黑色花岗岩、方钢、灰色砖、黑色砖、灰色壁纸、蓝色布艺等

Mashijiu POT is located in Aksu of the Xinjiang Uygur Autonomous Region, and it's a chain store which mainly serves Chinese food. Most cuisines focus on pot, which is very popular with people because of deliciousness. It's positioned for middle level dining market. This case of Chinese style presents a fashionable and simple Chinese restaurant in some active colors.

Main colors of this restaurant are grey, white and black. And the complementary color is lake blue. Designers adopt and combine these colors tactfully. Grey, white and black show Chinese classics, while blue makes the space lively. The combination of blue and white reflects fashion flavor. The pursuit of elegant and restrained eastern spirits is expressed very well. Meanwhile, more and more fashionable young people are fond of this restaurant, which let people feel Chinese culture from the aspect of modern outlook.

马仕玖煲位于新疆阿克苏，是一家主营汉餐的直营连锁店。主打以煲为主的菜品，美味口感，备受顾客好评。定位于中端餐饮市场。经过设计师与业主团队沟通，本案以中式为装饰风格，用出挑的色彩打造一个时尚、简洁的现代中式餐厅。

灰色、白色、黑色为本餐厅的主色调，湖蓝色为辅色，设计师巧妙地将这几种颜色穿插使用，融合在一起，灰白黑诉说着中式经典，而蓝色增添了餐厅的活跃气氛，蓝白结合凸显了时尚气息，在表达对清雅含蓄的东方式精神境界的追求的同时，也为更多时尚青年所喜爱，让人以现代的眼光感受中式的文化。

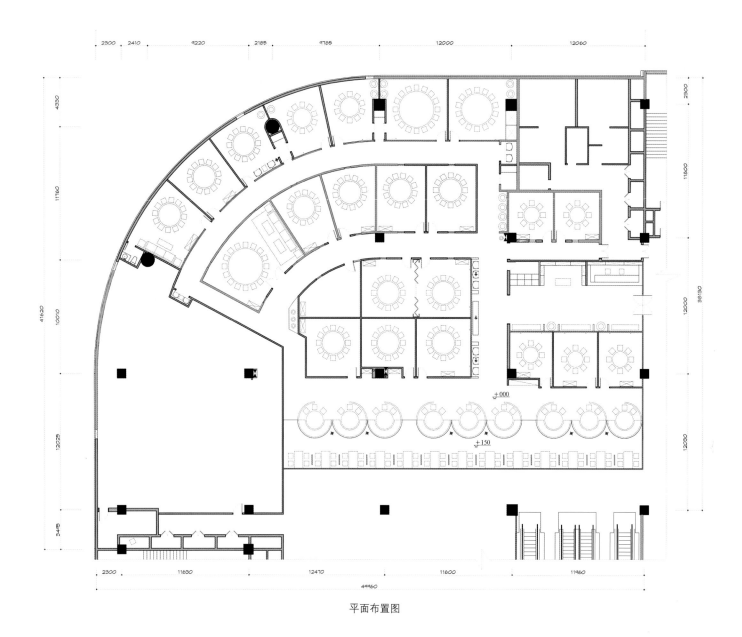

平面布置图

Functions of the restaurant are deployed quite clear. The whole moving lines of main and sub halls, table service areas and VIP rooms are very smooth. The main color of front hall at the entrance is black. Designers use light of VIP room tactfully and bring warmth through simple and smooth lined partitions. The floor is made of black granite which looks like mirror, echoing beveled glass in the ceiling, which extends the spatial feeling and makes people feel open and bright when stepping inside. Euphotic wooden latticed reception desk reflects the ground, which seems unique. There's a waiting area at the left side of front hall, with long Chinese stools whose function has been preserved and connected with surroundings, which also creates views when you walk around. Passing through front hall, you will be very impressed by the table service area. The round booths are shaped like bird cage made of white round tubes, while they won't block guests' vision and people can see the whole hall easily. The main color of table service area is white, matching blue cushion of sofa, which looks modern, fashionable, fresh and natural, and different from the grand atmosphere of front hall. Window frames are decorated with wooden batten partitions, presenting romantic atmosphere. The favorite part for lovers should be booths where they can enjoy the scenery outside the window. VIP rooms are next to table service area. Designers make good use of original space and divide them into rooms with different sizes, which can meet requirements from different guests and save space as well. The main color of VIP rooms is grey which can let people feel cozy. The original white painted ceiling seems special and keeps the height of VIP rooms. There're also delicate porcelain art paintings on the wall, simple Chinese chandeliers and exquisite ornaments form an elegant, simple and nice dining space.

This restaurant is not only a spatial design, but also a place where people inside can feel ancient eastern charm and modern fashion, which attracts you to enjoy its elegance and affection.

　　餐厅的功能布局分明，分前厅、散座大厅、包厢区，整体规划动线流畅。入口前厅以黑色为主色。设计师巧妙地运用包间的灯光，透过简单流畅的线条隔墙，暖意袭来，令人倍感温馨，而地面则用光面如镜的黑色花岗岩，与顶部的斜边条镜呼应，拉伸了空间感，使来客进厅倍感开阔。透光的木格前台倒影映入地面，别有风味。在前厅左侧设计了等候区，以长条中式凳为造型，人性化地保留功能部分，也与周边环境融合，一步一景。经过前厅，让你眼前一亮的是餐厅的散座区，映入眼帘的是以白色圆管组成的鸟笼圆形卡座。弧形的线条柔化了中式简练线条的生硬，半开放式的鸟笼隔断设计，不阻碍来客的视线，整个大厅尽收眼底。散座大厅以白色为主色调，蓝色的沙发垫做铺垫衬托，有别于前厅的大气，散座大厅整体表现得更为现代、时尚、清新自然。边窗用不规则的木条隔断墙，浓郁的浪漫气息扑面而来，赋予餐厅不一样的情调。拥有沿窗风景的卡座，应该是情侣顾客的最爱。绕过散座大厅则是餐厅包厢区，设计师利用原空间的结构，规划了大小不一的包间，分布合理，不浪费空间，也能迎合不同人数的客户群体。包间以灰色为基调，中性的颜色给人舒适感。原始漆白的裸顶，既新颖也保持了包间的高度，墙面精雕细琢的瓷盘艺术画，简约中式的主吊灯，一些精美的装饰品，形成格调高雅、造型简朴优美的就餐空间。

　　本餐厅不仅仅是一个空间设计，身临其境，来客感受到的是餐厅内蕴含着的古老华夏的神秘魅力与现代活跃的时尚氛围，使你不禁去细细品味它的韵味所在，情之所系。

Oulandi Café

欧兰迪咖啡厅

Designer: Jiang Guoxing
Design Company: Xupin Design Decoration Engineering Co., Ltd.
Project Location: Urumqi, Xinjiang
Building Area: 2,100 m²
Main Materials: Bronze paint, dark wood flooring,
　　　　　　　　black and white marble, black wood veneer,
　　　　　　　　mirror mosaics, tawny mirror, etc.

设 计 师：蒋国兴
设计公司：苏州叙品设计装饰工程有限公司
项目地点：新疆乌鲁木齐
建筑面积：2 100平方米
主要材料：古铜漆、深色实木地板、黑白根大理石、黑色木饰面、
　　　　　镜面马赛克、茶镜等

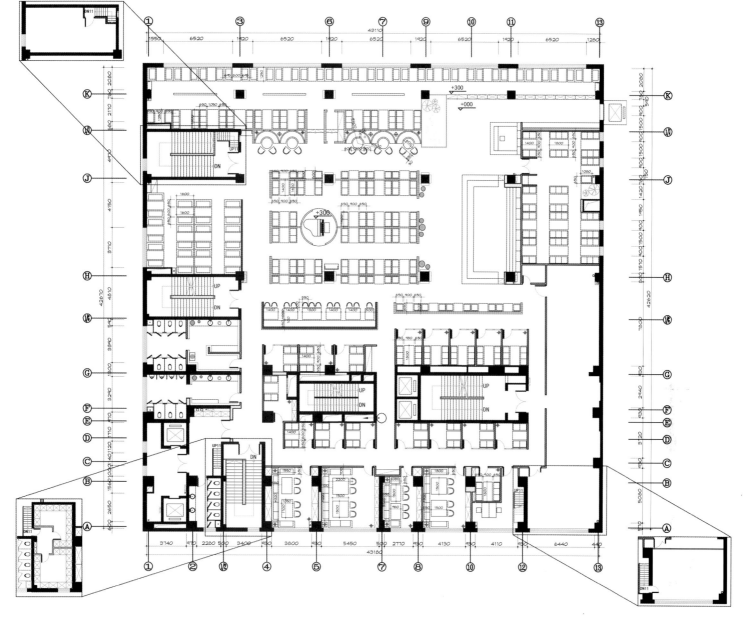

平面布置图

Oulandi Café is located in the prosperous business district in Urumqi of the Xinjiang Uygur Autonomous Region, and it's the biggest western food restaurant in Urumqi. The owner aims to develop the top brand in industry of western food in Xinjiang. After communications with designers again and again, the owner finally creates a unique and elegant western food restaurant.

The inspiration of this case comes from "lines". The designers make "the rhythm of lines" as theme of this case, and lines as main shape. Rhythm is the soul of space, coffee brown is main color of the space. Therefore, the requirements of structure and functions bring the rhythm which is likely to be controlled by symphony and has become the main factor affecting spatial aesthetics.

欧兰迪咖啡厅位于新疆乌鲁木齐繁华商业区，是目前乌鲁木齐面积最大的西餐厅。业主致力打造新疆西餐行业的顶尖品牌，经过与设计师的反复沟通、定位，一个独一无二、充满线条韵律的情调西餐厅就此亮相于众人眼前。

本案设计师的灵感来源于"线条"，将"线条的韵律"作为本案主题，以"线条"作为空间的造型，而韵律是空间的灵魂，深咖啡色为空间色彩的主旋律，韵律产生于结构与功能的需要，似乎由创作灵感所支配的交响乐谱曲那样受到控制，成为让空间美观的主要因素。

Layout of the restaurant is clear and smooth, and the function areas are divided clearly. There's a view corridor at the entrance, which also plays the role of waiting area. The corridor is next to the hall. European style shelf on the wall shows various kinds of art wares. Diamond solid ground with three colors, ceiling with mosaic mirror, greening at the end, and considerable special long stools for people to have a rest make people inside not feel bored. Walking through the view corridor, then you'll enter booth area. Designers adopt a lot of lines and bronze lines as partitions in the hall. Pillars are decorated by bronze lines which briefly introduce classical elegance and modern fashion in various ways and make the combination of spatial art and functions perfect. Windows seats can allow people to enjoy scenery outside through French windows decorated with patterns, and also bring light in. Indoor area employs main color of dark coffee, dotted by earth yellow and black. All of these elements make this café cozy and reflect restrained culture of this case.

餐厅的整体布局动线明朗、流畅，功能区域分布合理清楚，入口前厅是个兼具等候区功能的景观走廊，走道与大厅相邻，走廊侧墙面是整排摆放着工艺饰品的欧式展架，菱形三色穿插的立体地面，镜面马赛克的天花顶，与尽头景观绿化，并贴心设置别致长凳可供来客小憩，让来客在等候期间不会觉得枯燥。沿着景观过道行进便来到餐厅的散席大厅，设计师在处理大厅方面运用大量的线条装饰，古铜色的线条隔断，古铜色线条包饰的柱子，这些简练的线条，概括的基本造型语言，富于变化表情的粗细线条勾勒，既体现古典主义的韵味，又富有时代的气息，使整个空间艺术性和功能性结合得更加完美。沿边座位窗采用落地玻璃加花格，增加了观赏性，使顾客能更多地欣赏窗外风景，也加强了采光效果。内部空间以深咖啡色为主色调，以土黄色、黑色等为辅助色，营造出咖啡厅的舒适氛围，这些因素相融合，增加了餐厅所蕴含的文化气息。

Enjoying nice melody and tasteful coffee here when sunshine is shining on us, not only makes drinking coffee meaningful, but also creates spiritual pleasure, which encourages people to find inner peace in this cultural and cozy place.

置身其中，听着悠扬的背景音乐，点上一杯咖啡，透窗阳光洒在身上，享受的不仅仅是咖啡本身，更多的是精神的愉悦，让人远离烦琐、喧嚣的世俗，在这充满文化气息、舒适安逸的环境下找到属于自己心中的一片宁静。

Eastern Fence Narration

东篱·叙

Designer: Jiang Guoxing
Design Company: Xupin Design Decoration Engineering Co., Ltd.
Project Location: Urumqi, Xinjiang
Building Area: 1,200 m²
Main Materials: Wall cloth, dark solid wood flooring, dark wood veneer
dark wood-textured marble, volcanic rock,
cany wallpaper, grey mirror, ect.

设 计 师：蒋国兴
设计公司：苏州叙品设计装饰工程有限公司
项目地点：新疆乌鲁木齐
建筑面积：1 200平方米
主要材料：壁布、深色实木地板、深色木饰面、黑木纹大理石、火山岩、
藤制壁纸、灰镜等

The restaurant is located in the Nanhu Square of Urumuqi City, the Xinjiang Uygur Autonomous Region. The owner has more than ten years of experience in the restaurant business. With unique faith and insightful for dining, the owner positions the restaurant as a high-end Chinese food private club.

The designer prefers to treat it as an artwork rather than design it in a rigid style as the traditional Chinese restaurant. The combination of the quaint elements and fashion elements creates a new design conception. The warm soft hue of the restaurant and the whole design style offer customers a sense of serenity.

To insure the privacy of the restaurant, the designer puts an antechamber inside the entrance. The walls were covered with decorations made by twigs, and by the door are wooden chairs. The gurgling water in the circular crock adds fascination to the quiet space. Soft light falls on the niches. A sliding door blocks the view, offering boundless imagination to the customers.

There is an unimpressive entrance on the right. A row of microlandscape appears after entrance. The stairs on the right also block the view. The hidden entrance leads to the VIP box directly. This is so-called "A tortuous path leads to the quiet in the distance, where there is covered with flowers and trees shade the Buddhist temple". The same design is used in the hallway at the second floor. The designer makes the passage zigzag deliberately, and there are different visual attractions in every corner. At the end of the passage is a VIP box. The design reflected the aesthetics and retaines privacy at the same time.

The designer uses a lot of original furniture design in the dining rooms. The design makes an innovation on the Ming and Qing furniture, which will leave deep impression on customers. The use of original wood is not only environmentally friendly but also matches the design conception closely. This is the goal of both the designer and the owner.

Using the cany wallpaper will change customers' impression of restaurant rooms. The cane is glossy, elastic and smooth to touch. Such kind of wallpaper is an excellent material for decoration, which creates the peaceful atmosphere as well as the new aesthetics. Most of the lamps are in arc shape, which reflects the traditional conception that "the Earth was square and Heaven was round". This creates smooth permeability for the space and accords with the overall style.

Several boxes and passage are separated by black wood lattices, which make the restaurant more active and permeable. The paintings, the clay pots and the books also increase artistic conception to the space.

With the higher cognition degree of cuisine art, customers taste the food as well as the environment. Mr. Tao's attention is not on the cup but on the atmosphere. The mood is partially decided by the environment. The designer interpreted the conception perfectly.

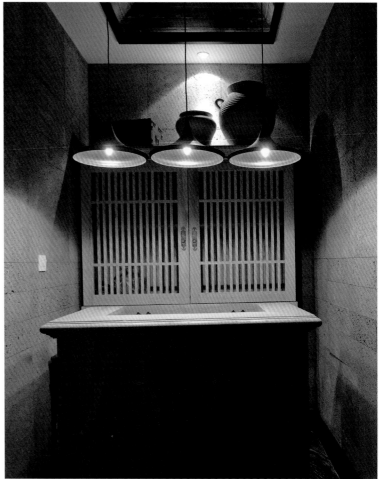

东篱·叙餐厅会所位于新疆乌鲁木齐南湖广场，业主是个有着十多年餐饮经验，对于餐饮有着自己的信念和理解的人，该餐厅定位在高级中餐厅私人会所，经营中餐。

设计师不拘泥于传统的中式餐厅的严谨肃静，更愿意把它当作一个艺术品来对待，运用中国传统古朴的设计元素并且推陈出新，创造新的设计元素，带来新的设计理念，加之餐厅柔和温暖的主色调，让人心里顿生一种皈依的宁静感。

入口处，设计师考虑了餐厅的私密性。入口进去是一个小前厅，门口放着古朴的木椅子，墙上是原生态的树枝挂饰，旁边有个圆形的瓦缸，淙淙的水流让安静的空间显得更加神秘，墙上有几个壁龛，柔和的灯光投在粗犷的陶器上，一扇推拉门挡住了视线，让人对门后的风景产生无限的联想。

入口处的右边还有一个容易被人忽视的小入口，进去之后映入眼帘的是一排拙朴的小景观，右边转过去，一段楼梯遮住了视线，走上去，转一个弯，再走一段才到二楼，这个隐蔽的入口直接通往VIP包厢，像是走在一条幽静的小路，来到一个私人的小屋，这就是所谓的"曲径通幽处，禅房花木深"。同样的设计也用在了二楼的走道，从平面上看，设计师也是故意把通道做成折线形状，运用了园林的设计手法，在客户路过的每个转角都会有不同的视觉景点，步步皆景，最后到达一个VIP包厢。这既体现了美观性，又保留了私密性、神秘感。

餐厅包间使用了大量的原创性家具设计，设计方面在明清家具的基础上做出了创新。这一独特的设计，相信会给人留下深刻的印象。家具采用原生态的木材，不仅环保并且与餐厅的设计理念"回归自然，学会生活"相吻合。这正是设计师和业主想要达到的结果。

包间多采用藤制壁纸，改变以往人们对餐厅包间的印象。藤条外皮色泽光润，手感平滑，弹性极佳，色彩柔和，对于包间墙面装饰是极佳的材料。不仅使包间处于一种宁静的氛围中而且创造了一种全新的美感和质感。包间灯具造型大多采用柔和的圆弧，寓意中国传统中"天圆地方"的博大精深的思想，为空间制造一种流畅的通透感，与整个餐厅整体温馨的格调一致。

其中几个包间和过道采用黑色木格划分空间，使得整个餐厅不会显得呆板沉闷，与大面积的藤制壁纸形成对比，增加了餐厅整体空间的通透感、趣味性。在软装方面，墙壁挂的水墨画，柜子上摆放的陶罐和书籍，整体统一，丰富了餐厅空间，使得餐厅空间有了书香、禅意的味道。

如今，随着人们对餐饮艺术的认知度越来越高，品位也越来越高。美食，不仅仅是舌尖上的享受，更是环境氛围的享受。就像陶渊明先生一样，意不在酒，在乎的是环境、氛围，在什么样的环境下品尝才能有什么样的心境。设计师将以上所有的设计理念已经诠释得非常完美。

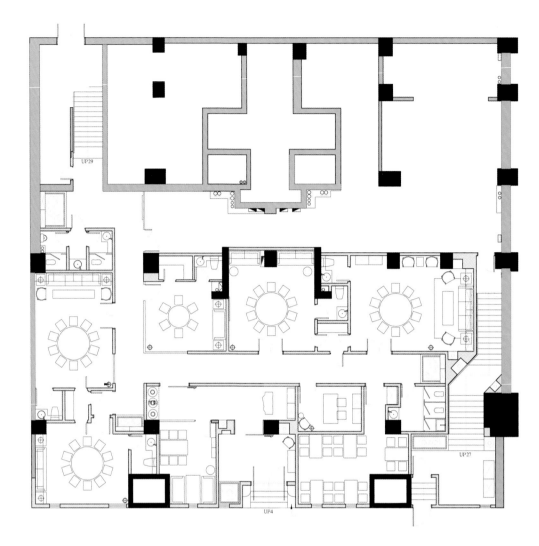

一层平面布置图

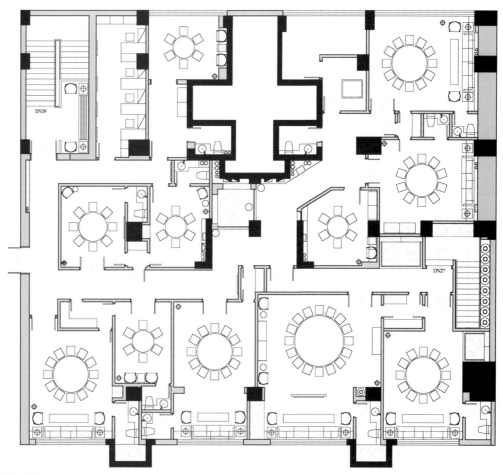

二层平面布置图

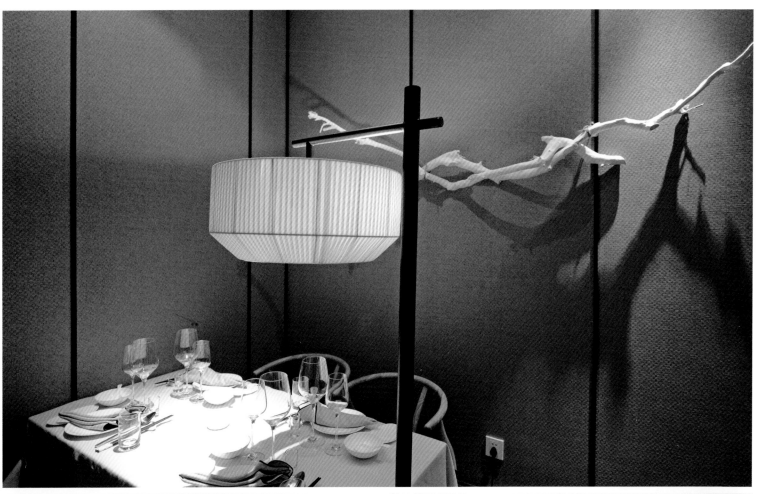

255

Xupin Design Xinjiang Branch Office (Times Square)

叙品设计新疆分公司（时代广场）

Designer: Jiang Guoxing
Design Company: Xupin Design Decoration Engineering Co., Ltd.
Project Location: Xinjiang
Building Area: 300 ㎡
Main Materials: Chinese black mirror marble bricks, black wood flooring, White of Volakas, straw paint, soil paint, square steel, wire netting, mirror, etc.

设 计 师：蒋国兴
设计公司：苏州叙品设计装饰工程有限公司
项目地点：新疆
建筑面积：300平方米
主要材料：中国黑镜面处理大理石砖、黑色木地板、爵士白大理石、稻草漆、泥土色漆、方钢、铁丝网、镜子等

Xupin Design Xinjiang Branch Office is a new works done by the designer from 2013 to 2014.

Huge changes have happened to modern office in last decade. Various kinds of designs are used to offices. Designer of this case combines modern and traditional elements after getting the acknowledge of traditional culture, creating a classic charming office according to the requirement on aesthetics of modern people. The original and novel design has totally made people change their thoughts about modern office.

Black is the main tone of this case. The simple hard straight lines show mature design methods of the designer. Chinese square steel partition at the first floor can not only separate the functional areas, but also let each area connected but not blocked. There're mainly leisure and fashionable sofa area, working area for dozens of people and an area of service counter at the first floor. These three functional areas seem separate but one unit. In order to pursue a natural and close feeling, the designer uses soil paint in the original ceiling, and straw paint on the walls. All these inspirations are unique and cannot be copied.

A mirror image space is created at the second floor. Mirror and wire netting can bring special feelings. The feature of reflection of the mirror is used in partitions, which makes originally small space look more spacious and prospective. The second floor is mainly used for working, meeting and receiving visitors. Bar section in meeting area can not only allow you to drink wine or tea, but also creates a comfortable communication atmosphere. Walking upstairs, you will find there're two Chinese round chairs along the aisle, which brings eastern Buddhism elegancy. Round window of the designer's office borrows the perspective design of Chinese garden.

All funishings of this case are guided by the designer, and it's also the teamwork with other designers. Therefore, as for those employees working in this office, it's a life and design experience.

With traditional culture as soul, this case gets rid of the accumulation of symbols in modern decorative way. The designer pays more attention to the creation of spatial artistic conception, implantation of natural landscape and involvement of emotion, and focuses on the control of style. Simple traditional Chinese elements are kept, meanwhile, eastern artistic conception are reserved, which makes the whole office be in a cozy and natural atomosphere.

叙品设计新疆分公司——时代广场A座25层-H（新址），是设计师2013年至2014年又一项新的力作。

现代办公在过去的十几年里发生了巨大的变化，各种设计风格的办公室层出不穷。本案中设计师通过对传统文化的认识，将现代元素和传统元素结合在一起，以现代人的审美需求来打造富有传统韵味的办公空间。这种原创性的、新颖的设计完全地改变了以往人们对现代办公室的认识。

本案中以黑色为主色调，简洁硬朗的直线条，体现了设计师既简练又熟练的设计手法。其中一楼中式的工艺方钢隔断既能把每个功能区域划分开又可以让每个区域可以互相交流不至于很闭塞。一楼主要为休闲、时尚的会客沙发区、可供十几人办公的办公区和一个服务台区。这三个区域看上去分开，但又是一个整体。为了追求一种自然、亲切的感觉，分别在原顶上喷了泥土颜色的漆，在墙面上喷了稻草漆。这些设计灵感都是独特的、不可复制的。

二楼主要创造了一个镜像空间，突出镜子和铁丝网材质给人的特别感受，将镜子反射的特点运用在隔断上，使本来很小的空间看起来更加宽广，更有意境。二楼集合了设计师办公、会议、会客这些功能。会客酒吧区可以品酒喝茶，又可以舒适地交谈。从楼梯上去，两把中式圈椅摆在过道边上，立刻产生一种东方禅意韵味。设计师办公室中圆窗的设计，主要借鉴了园林中透景的设计手法。

本案中所有的软装由主案设计师本人直接参与指导，在其他设计师的共同努力下完成所有装饰。所以在本案对公司员工来说，更是一种生活体验和设计体验和情感的投入。

本次案子的设计以传统文化为灵魂，以现代装饰为手法，摒除了符号的堆砌。设计师特别注重空间意境的营造、自然景观的植入、情感的注入，着重于控制空间的品位，在精简传统中式元素的同时，又不失东方意境，使得整个设计呈现一种舒适、自然的办公氛围。

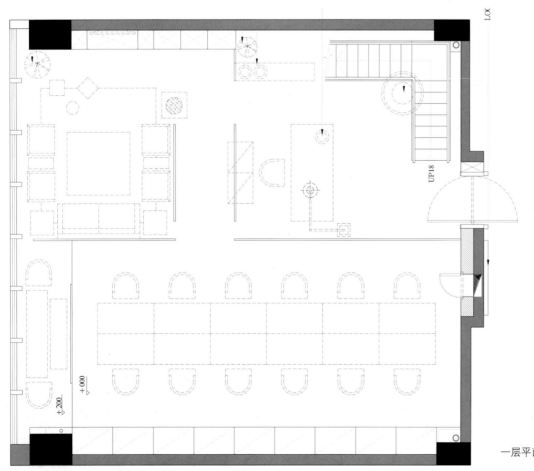

一层平面布置图

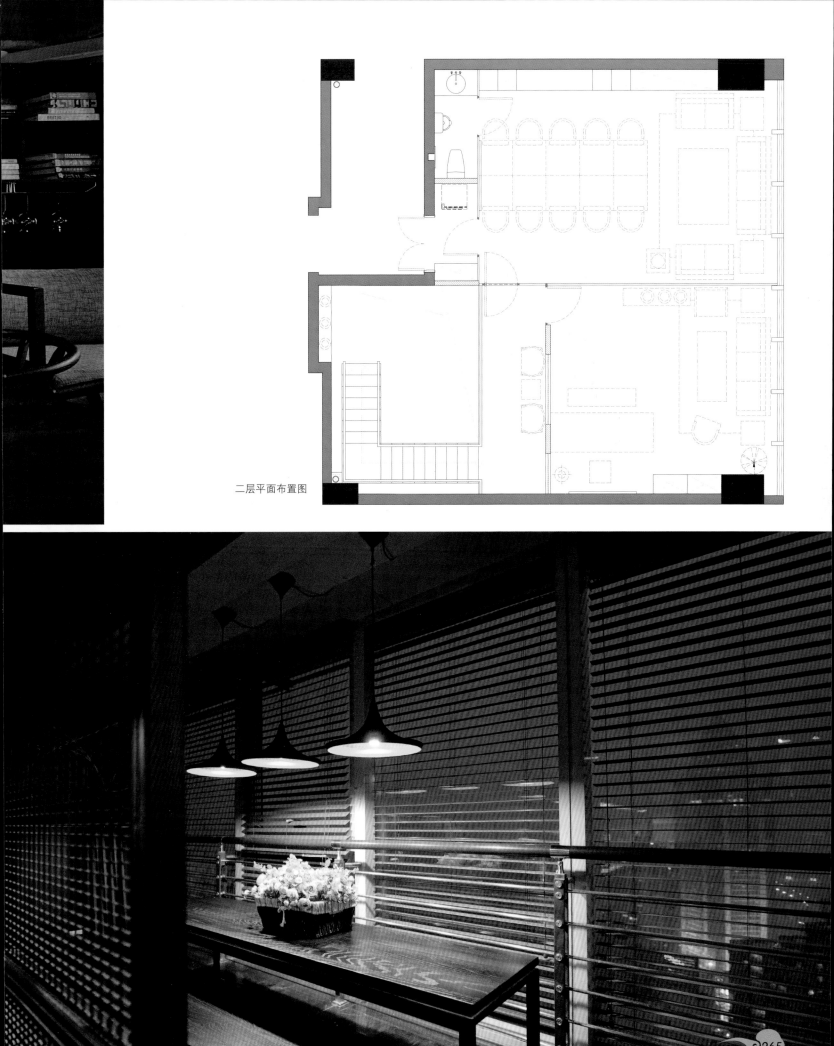

二层平面布置图

Xupin Design Shandong Branch

叙品设计山东分公司

Designer: Jiang Guoxing
Design Company: Xupin Design Decoration Engineering Co., Ltd.
Building Area: 650 ㎡
Main Materials: Light gray wallpaper, dark wooden flooring, black square steel, Chinese hollow wood lattice, black wood veneer, etc.

设 计 师：蒋国兴
设计公司：苏州叙品设计装饰工程有限公司
建筑面积：650平方米
主要材料：浅灰色壁纸、深色木地板、黑色方钢、中式镂空木格、黑色木饰面等

This case is the third branch of Xupin Design, located in Global International Building Linyi City Shandong Province. The office is a collective space for brain work, while the designers' office is not just limited to the pure office, because its environment has an impact on the designers' creation and also expresses the design philosophy of the company and the passion and dedication for the design, it has to allow the customers to directly feel the company's creativity and the design value, giving customers a sense of trust in the company.

Designers focus on the traditional Chinese spirits in the case and adopt simple approaches to modify the entire space. The core color of light gray has experienced a gradual change from shallow to deep. Dotted with green plants, it creates an introverted and comfortable atmosphere full of tension. The entrance is arranged with the reception desk and the LOGO backdrop to greet visitors. The designers use the irregular interspersion of the original concrete wall and black square tubes to create the big LOGO "Xupin" which emanates an introverted and hazy feeling. This simple greeting space is paved with dynamic wooden floorings, which is used as the buffer zone connecting with various areas of the company. In its sides, designers apply Chinese wood lattice and bookcases to separate the space from others so that the space has a good permeability and overall lighting so as to make the visitors feel the space's transparency and generosity. The design department, conference room, reception area, and drawings discussing are placed laid in both sides of the reception desk. The delicate spatial layout invisibly lets people refreshing. In the side of the design department, there is an exhibition space to mainly display the company's excellent works over the years. Walking along the path in the exhibition space, visitors then come to the built-in lounge, unconsciously feel the company's development history, and get a deep understanding of the company.

Novel black birdcage lamps, exquisite Chinese wood lattices, calm and introverted Chinese furniture, bookcases filled with books, etc., all of these together create a stylish Chinese office space which is not limited to the traditional Chinese style yet with the traditional connotation, and contains a little bit of Zen artistic conception and the vitality in the quietness.

本案是叙品设计公司的第三个分公司，位于山东省临沂市环球国际大厦。办公室是脑力劳动的集体空间，而设计师的办公室不仅仅限于纯粹的办公，它的环境影响设计师的创作，也直观地体现设计公司的设计理念和对设计的热情及执着，让往来的客户直观感受公司的创造力与设计的价值观，赋予客户对设计公司的信任感。

本案设计师注重传统的中式精神意境，延伸简练而不简单的手法修饰整个空间，浅灰色作为核心色彩基调，经历了由浅至深的渐变，加以绿色植物的点缀演绎出一种内敛、安逸、富于张力的空间情调。入口，迎接访客的是公司前台及LOGO背景墙，在此运用原混凝土墙及黑色方管不规则的穿插，构成的"叙品"大LOGO，于朦胧中透着一丝内敛。简约的前台运用动感十足的条形木地板，前台空间连接着公司的各个区域，作为空间的缓冲带。在它侧面，设计师运用中式木格及书柜划分区域空间，这样让空间具有通透性，又不影响整体采光，令访客感觉空间通透大气。设计部、会议室、会客区、图纸讨论区合理地安排在前台两侧，精致的空间布置，使人身心舒畅。在设计部的侧面是访客可以欣赏的展示空间，主要展示公司这些年来的优秀作品。访客沿着展厅走廊的过道再到内置的休闲室，不知不觉中感受到公司的发展足迹，更深层次地了解该公司。

新颖的黑色鸟笼灯、玲珑的中式木格、沉稳内敛的中式家具、布满书籍的书柜等等，这些元素相结合营造出一个时尚与中式相互渗透，不拘泥于传统中式的条框，又不失传统的内涵，带着点点禅意，宁静中充满着活力的办公空间。

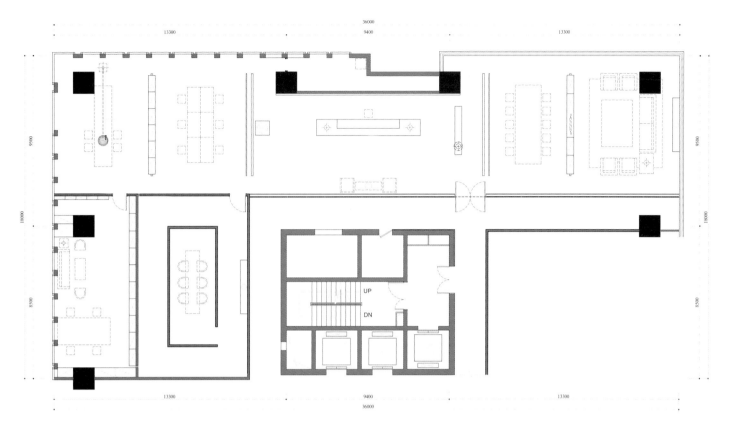

平面布置图

Xupin Design (Zhenchuan Road)

叙品设计（震川路店）

Designer: Jiang Guoxing
Design Company: Xupin Design Decoration Engineering Co., Ltd.
Building Area: 1,000 ㎡
Main Materials: Mushroom stone, carpet, wood batten, wallpaper, laminate flooring, glass, hollow bricks, white latex paint

设 计 师：蒋国兴
设计公司：苏州叙品设计装饰工程有限公司
建筑面积：1 000平方米
主要材料：蘑菇石、地毯、木条、墙纸、复合地板、玻璃、空心砖、白色乳胶漆

In the case, the colors and the layout are different from the traditional approach. Through the designer's skillful matching white with other colors, white shows a deep artistic conception.

Borrowing the technique of exalting after abasing in the traditional garden design, the designer makes use of the low-key doors, narrow corridors, and the hidden entrance to achieve an effect of being in an open and amazing space in the turning, with the embellishment of cobblestone, giving out a refreshing feeling. Going through the corridor, you will slowly forget earthly troubles.

空间设计中，在色彩和布局上，都跳脱传统，独树一帜。巧妙的白色运用以及其他颜色的配合，呈现了"此时无声胜有声"的意境。

借用了传统园林设计中欲扬先抑的手法，低调的门、狭窄的走廊、隐蔽的入口，却在转身的一刹那豁然开朗，别有洞天。鹅卵石夹道，清新怡人，走过长廊，尘世的烦恼也慢慢抛诸脑后。

Green symbolizes life, and white stands for elegance. From the fixture to the furnishings, colors are harmonious and unified, which reflect the perfect pursuit of the details. People and objects coexist here and blend together.

The theme of illusion makes a good modern interpretation of Chinese culture, so the space is always emanating an unexpected wonderful feeling at the ting places of the pure and simple style.

The office is a place where people are engaged in brain work, while the emotions and working efficiency of the staff will often be affected by the environment, so this new office pays more attention to the cheerful and pleasant colors and chic creative ideas in the design. Immersed in with the fragrance of tea in the air, the staff can complete their work in a relaxed mood and also improve their work efficiency.

This is not only a healthy way of working, but also reflects people's positive attitude towards life. In this materialistic age, the case is designed to let irritable people refind the original modesty so as to calm down to work happily and feel the real life.

　　绿色象征生命，白色象征优雅，从硬装到家具陈设，配色都协调统一，体现对细节的完美追求。人和物在这里共存，融为一体。

　　"梦幻"的主旨对中式文化进行很好现代化表现，空间总会在细小处给你意想不到美妙感受，简洁纯净的主题始终贯穿整个案子。

　　办公室是人们从事脑力劳动的场所，员工的情绪、工作效率常常会受到来自环境的影响。而在叙品公司的这间新办公室中，轻松愉快的色彩、巧妙的创意，再加上空气中弥漫的茶香味，所有这些都可以让员工在放松的心情下完成工作，从而有利于促进工作效率的提高。

　　这不仅是一种健康的工作方式，更体现了人们积极的生活态度。在这物欲横流的年代，本案的设计希望让烦躁的人们重新寻找到最初的那种质朴，静下心来快乐地工作，同时能感受真正的生活。

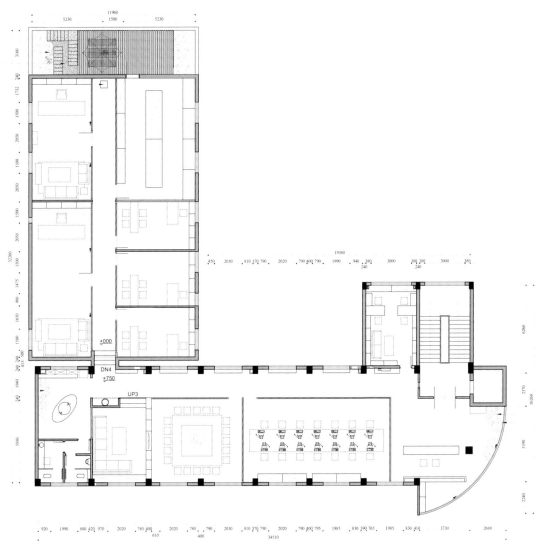

平面布置图

Longhai Construction

龙海建工

Designer: Jiang Guoxing
Design Company: Xupin Design Decoration Engineering Co., Ltd.
Project Location: Kunshan, Jiangsu
Building Area: 1,800 m²
Main Materials: Black granite (mirror finish), dark wooden flooring, light gray wallpaper, gray wall tiles, black wood veneer, etc.
Photos: Jiang Guoxing

设 计 师：蒋国兴
设计公司：苏州叙品设计装饰工程有限公司
项目地点：江苏昆山
建筑面积：1 800平方米
主要材料：黑色花岗岩（镜面处理）、深色木地板、浅灰色壁纸、
　　　　　灰色墙面砖、黑色木饰面等
项目摄影：蒋国兴

The office of Longhai Construction is located in Dong'an Building Kunshan Jiangsu. Kunshan is a county-level city with the strongest economic strength. The rapid economic development in recent years has attracted many enterprises to vest, and Longhai Construction is one of them.

In the case, the designer uses a balanced symmetric principle to plan out a smooth and vibrant space. The design is the combination of Chinese traditional culture and elements in a modern technique so that the overall space is dignified and elegant. The background of the reception desk in the hall is decorated with stones which are specially selected from the production place of origin, picturesque and elegant, which is matched with the waterscape and the restrained rustic Chinese furniture, emanating a special charm, giving people rich reveries and revealing the beauty of Chinese mood.

　　龙海建工办公室位于江苏昆山东安大厦。昆山是国内经济实力最强的县级市之一。由于近几年经济的快速发展，昆山吸引了更多的企业入住，龙海建工便是其中之一。

　　本案设计师运用均衡对称的原则来规划布局，使整个空间流畅明快。整个办公室以现代手法将设计师所理解的中式传统文化与中式元素相融合，整体庄重、淡雅。大厅的前台背景特地在石材产地挑选，它如墨如画，淡雅出尘，妙在似与不似之间，再配以旁边的水景以及内敛、质朴的中式家具，让人觉得别有一番韵味，使人产生丰富的遐想，无形中透露出中式的意境之美。

The aisle adopts simple and tough lines to sketch out the layers. The plain techniques of the virtual-real comparison and the change of light and dark are employed to create a simple space. In the turning, there is a row of white and elegant birches space with straight trunks. The tough, upright, honest spirits of the birch is just a symbol of the corporate culture.

The internal space is placed with large bookshelves loaded with books and the racks for the display of porcelains, which silently tells the owners' cultural literacy. And the ceiling adopts the design of the rectangle floating top, which brings a visual impact through the collision with the lighting, so that the tranquil atmosphere adds a touch of the active and open mind. In addition, there is another scene in the balcony of the atrium which consists of wood preservative, wall tiles and natural plants, making a chance to get close to the nature and creating a casual atmosphere.

过道用简洁硬朗的直线条勾勒层次感，以朴实的手法，通过虚与实、明与暗之间的对比营造简约不简单的空间。拐弯处设置一排树干修直，洁白雅致的白桦树，它坚韧、挺拔、正直的精神是企业文化的一种象征。

办公室内部运用大面积的满载书籍的书架和摆放瓷器的饰品架，这些无声诉说着办公室主人的文化素养。顶部则用简练的矩形浮面来体现，与灯光的碰撞让你在视觉上受到冲击，为这宁静安逸的空间增添了活跃气氛，利于开阔思维。公司中庭阳台处还有另一番景象，它这用了防腐木、墙面砖及自然植物景观，在此人可以与自然亲密接触，享受这休闲氛围。

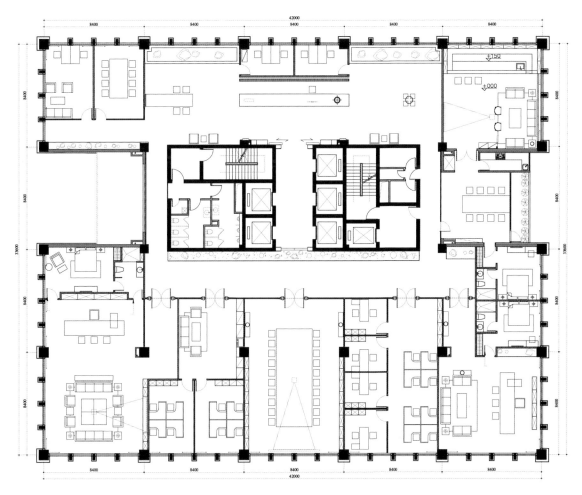

平面布置图

图书在版编目(CIP)数据

蒋国兴作品集 / 蒋国兴主编. －－ 南京：江苏科学技术出版社, 2014.6
 ISBN 978-7-5537-3077-6

Ⅰ.①蒋… Ⅱ.①蒋… Ⅲ.①室内装饰设计－作品集－中国－现代 Ⅳ.①TU238

中国版本图书馆CIP数据核字(2014)第081085号

蒋国兴作品集

主　　　编	蒋国兴
项 目 策 划	徐宾宾
责 任 编 辑	刘屹立
特 约 编 辑	赵　萌

出 版 发 行	凤凰出版传媒股份有限公司
	江苏科学技术出版社
出版社地址	南京市湖南路1号A楼，邮编：210009
出版社网址	http://www.pspress.cn
总 经 销	天津凤凰空间文化传媒有限公司
总经销网址	http://www.ifengspace.cn
经　　　销	全国新华书店
印　　　刷	利丰雅高（深圳）印刷有限公司
开　　　本	1 020 mm×1 440 mm　1/16
印　　　张	19
字　　　数	152 000
版　　　次	2014年06月第1版
印　　　次	2014年06月第1次印刷
标 准 书 号	ISBN 978-7-5537-3077-6
定　　　价	298.00元（精）

图书如有印装质量问题，可随时向销售部调换（电话：022-87893668）。